D1029336

REGULATED ENTERPRISE

HISTORICAL PERSPECTIVES ON BUSINESS ENTERPRISE SERIES
Mansel G. Blackford and K. Austin Kerr, Editors

Managing Industrial Decline: The British Coal Industry between the Wars
Michael Dintenfass

Henry E. Huntington and the Creation of Southern California
William B. Friedricks

Making Iron and Steel: Independent Mills in Pittsburgh, 1820–1920
John N. Ingham

Eagle-Picher Industries: Strategies for Survival in the Industrial
Marketplace, 1840–1980
Douglas Knerr

A Mental Revolution: Scientific Management since Taylor
Edited by Daniel Nelson

Rebuilding Cleveland: The Cleveland Foundation and Its Evolving
Urban Strategy
Diana Tittle

Daniel Willard and Progressive Management on the
Baltimore & Ohio Railroad
David M. Vrooman

Regulated Enterprise

*Natural Gas Pipelines and
Northeastern Markets, 1938–1954*

Christopher James Castaneda

Ohio State University Press   Columbus

Library of Congress Cataloging-in-Publication Data

Castaneda, Christopher James, 1959–
    Regulated enterprise: natural gas pipelines and northeastern markets, 1938–1954 /
Christopher James Castaneda.
        p.   cm. —(Historical perspectives on business enterprise series)
    Includes bibliographical references and index.
    ISBN 0–8142–0590–9
    1. Gas industry—Government policy—Northeastern States—History—20th century.
    2. Natural gas pipelines—Northeastern States—History—20th century.
    I. Title. II. Series.
    HD9581.U52A1153 1993
    388.5'6'0974—dc20   92–23827
                                                                              CIP

Text and jacket design by Jim Mennick.
Type set in Times Roman by Focus Graphics, St. Louis, MO.
Printed by McNaughton & Gunn, Saline, MI.

9 8 7 6 5 4 3 2 1

In memory of Terrill Lynn Castaneda

Contents

Maps

Tables

Acknowledgments

IN THE preparation of this work, Dr. Joseph A. Pratt offered invaluable assistance as mentor, graduate adviser, and friend. In December 1985 during my first conversation with him he introduced me to the topic of the transition from manufactured to natural gas by northeastern utilities as a possible case study that I would do for him as a research assistant. The resulting brief study gradually evolved into a dissertation on the regulated pipelines which supplied the natural gas to those northeastern utilities. Dr. Pratt also provided me with several research opportunities in corporate history projects at Texas Eastern Corporation (now a subsidiary of Panhandle Eastern Corporation) and Tenneco, Incorporated. Participation in these projects allowed me to greatly enhance the quality of this work; I also express my appreciation to those two firms, which opened their records. In addition, I thank the Business History Group (Dr. Louis Galambos, Dr. Joseph Pratt, and Robert Lewis) for generous support in writing the dissertation on which this book is based. Tab Lewis, T. Lane Moore, and William Creech of the National Archives directed me to useful records. I also received welcome encouragement from many faculty members and graduate students at the University of Houston, where I completed my graduate studies and revised this work.

Those who read the manuscript at various stages of development offered useful suggestions. I owe much appreciation to my editor, Mansel B. Blackford, for both reading the manuscript and offering insightful suggestions on expanding its theoretical framework; an anonymous reader also provided significant suggestions for improving the manuscript. In this light, I offer thanks to Alex Holzman of the Ohio State University Press for making contact with me at a business history conference regarding my dissertation and suggesting I keep in touch with him about it—which I did. Additional readers included Alan D. Anderson, James A. Castaneda, Kenneth Lipartito, Martin Melosi, Joe Pratt, and Alan Stone.

Others assisted in a variety of ways. Fran Dressman, Catherine Felsmann, Dr. John O. King, Bob Kirtner, Trey Mecom, David Munday, Dr. Robert Palmer, Brad Raley, Clarance Smith, and Christine Womack deserve mention for a diversity of important reasons, not the least of which was friendship during the course of this work.

Finally, I express gratitude to my wife, Terri, for her support and encouragement and to my children, Courtney and Ramsey, just for being there.

1. Introduction

THE INTERSTATE natural gas pipeline industry is an important and largely unstudied topic for students of business and government relations. As a highly regulated capital-intensive business, the pipeline industry is a useful study for better understanding how government and business interaction in a democratic society enables or hinders responsiveness to market forces. In this particular work covering a major episode between 1938 and 1954 which occurred during the first sixteen years of federal regulation of the interstate gas pipeline industry, it is apparent that the regulatory process fostered competitive growth in the natural gas industry. During these years, the percentage of natural gas consumed in the United States compared to other fuels rose from approximately 11 to 26 percent (see table 1.1).

Federal regulation of the industry began with the provisions of the Natural Gas Act (1938). The act authorized the Federal Power Commission (FPC) to regulate the interstate transmission and sale of natural gas. In 1954 the United States Supreme Court extended the FPC's regulatory jurisdiction to the price of gas produced for interstate commerce. Scholars have examined in detail post-1954 producer regulation, not pre-1954 federal pipeline regulation. Thus, existing business and government relations scholarship ignores an extremely important era of federal regulation.

From 1938 to 1954, the FPC's primary jurisdiction was the regulation of the gas pipeline industry through the issuance of certificates of public convenience and necessity. Intense competition between pipeline companies and other interests characterized one significant episode in this era as entrepreneurs formed gas lines and vied to acquire certificates authorizing the transmission of southwestern-produced natural gas to the

Table 1.1. United States Energy Consumption by Fuel, 1920–1970
(percentages)

Year	Coal[1]	Crude Oil[2]	Natural Gas[3]	Water Power	Nuclear	Total
1920	78.4	13.3	4.4	3.9	—	100.0
1925	70.4	19.9	6.4	3.3	—	100.0
1930	61.2	25.4	9.9	3.5	—	100.0
1935	55.7	28.8	11.2	4.3	—	100.0
1940	52.4	31.4	12.4	3.8	—	100.0
1945	50.7	30.5	14.1	4.7	—	100.0
1950	37.8	37.2	20.3	4.7	—	100.0
1955	29.3	40.8	26.1	3.8	—	100.0
1960	23.2	41.6	31.6	3.6	—	100.0
1965	22.3	40.1	33.7	3.8	0.1	100.0
1970	18.9	40.3	36.5	4.0	0.3	100.0

NOTES: Figures do not account for energy provided by wood.
[1]Includes bituminous coal, lignite, and anthracite.
[2]Adjusted for net exports or imports of petroleum products.
[3]Includes natural gas liquids.
SOURCE: Library of Congress, Congressional Research Service, *The Energy Factbook* (Washington, DC: GPO, 1980), 13.

major metropolitan areas of the Northeast, which had no access to natural gas. The FPC provided an effective forum for the expansion of the gas pipeline industry from southwestern supply into new northeastern markets.[1]

REGULATION OF THE HISTORICAL NATURAL GAS INDUSTRY

The history of the American natural gas industry can be divided into four major time periods as defined by varying levels of federal regulatory involvement. The first era (1816–1937) began with the incorporation of the nation's first natural gas company in 1816. It was characterized by a total lack of federal involvement. Regulation was left to the jurisdiction of state and local commissions, which had virtually no authority to regulate gas companies operating in interstate commerce.[2] By the late 1920s, the burgeoning interstate gas pipeline industry generally escaped all regulation and contributed to the larger problems created by the few huge public utility holding companies which controlled much of the nation's energy.

The Natural Gas Act (1938) opened the second phase of development in the industry, which lasted into 1954. The act empowered the FPC to regulate the interstate transmission and sale of natural gas through the certificate of public convenience and necessity. Without such a certificate, gained normally after lengthy and formal public hearings before the FPC, a pipeline company could not operate across state lines. During this period, according to M. Elizabeth Sanders, "the relationships between consumers, regulators, and business were generally harmonious."[3] The Natural Gas Act, however, was poorly worded in certain sections, and disputes of interpretation over these sections, particularly in regard to the regulation of gas production, created substantial problems for both Congress and the FPC.

The Supreme Court's so-called Phillips decision of 1954 answered this dispute and marked the beginning of a new era of increased federal involvement and business-government acrimony in the natural gas industry; this is the period most commonly studied by scholars.[4] The Phillips decision substantially raised the FPC's powers by giving it the authority to regulate the wellhead price of natural gas sold by producers to pipelines for interstate commerce. The FPC's subsequent strict control of wellhead prices led to a severe natural gas shortage in the early to mid-1970s. The resulting gas crisis, part of the nationwide energy crisis of that decade, resulted in a renewed effort to deregulate the industry.[5]

The Natural Gas Policy Act of 1978 (NGPA) launched the industry's fourth era. The Federal Energy Regulatory Commission (FERC), the FPC's successor, implemented a new wellhead-gas-pricing system intended to gradually deregulate the gas-producing industry. Since 1978, there have been numerous efforts designed to further deregulate the industry and transform gas pipelines from their historical merchant function, buying and selling gas, to common carrier status.

AN OVERVIEW: COMPETITION FOR NEW MARKETS, 1938–1954

Although the history of the pre-1954 gas industry is essentially missing from scholarly literature, the period 1938 to 1954 contains a fascinating episode of regulated competition for new markets. Why, then, is this a virtually unknown story? A partial answer is that this period also involved an overshadowing congressional debate on viability of the Natural Gas Act generally, and its jurisdiction over gas producers, in particular; several scholars have studied these deliberations in detail.[6] This era, however, is clearly characterized by rapid industry growth in several markets throughout the nation. The single most important example involved the creation of several pipeline companies that competed with one another and against the coal and railroad industries to acquire certificates to sell southwestern natural gas in northeastern markets.

The Northeast represented a vast market for natural gas sales. In Appalachia, declining regional production required new supplies to meet demand. But farther to the northeast, cities such as Philadelphia, New York, and Boston remained dependent upon manufactured gas, an inefficient and increasingly expensive gas. The coal industry, which supplied the raw product for manufactured gas, and the railroad industry, which transported the coal, adamantly opposed but ineffectively resisted the introduction of natural gas into their traditional northeastern markets. Coal and railroad interests attempted to hinder and delay the expanding gas industry while coal and manufactured gas continued to lose sales to natural gas (see table 1.2).

In the Southwest, aggressive and politically sensitive entrepreneurs formed and guided three companies to sell the massive quantities of previously unmarketed natural gas to northeastern consumers. Tennessee Gas Transmission Company, Texas Eastern Transmission Corporation, and Transcontinental Gas Pipe Line Corporation vied for northeastern gas customers.[7] Entrepreneurs and managers including Curtis Dall and Gardiner Symonds at Tennessee Gas, E. Holley Poe, Reginald Hargrove, and George Brown at Texas Eastern, and Claude Williams at Transcontinental exploited opportunities in both the gas industry and the regulatory environment to build and operate these lines.[8] The three pipelines eventually introduced natural gas into the Northeast on a scale large enough to ensure a long-lasting and generally stable supply for the entire region while simultaneously providing a market for voluminous quantities of Gulf Coast natural gas.[9]

The initial impetus for expansion came during World War II. Vital war production factories in Appalachia required more natural gas than local producers could supply. In response to this business opportunity, a small group of entrepreneurs organized Tennessee Gas. But difficulties in arranging financing and gas supply during wartime forced the original promoters to sell their company to an established corporation. The new owners quickly built the line with essential assistance from both the Reconstruction Finance Corporation and the War Production Board (WPB).

After World War II, in a continuing episode of federal war-agency involvement in the gas industry, the War Assets Administration auctioned in a highly politicized bidding competition two war emergency petroleum

Table 1.2. Gas Sales of Utilities in the Northeast and the United States,[1] 1935–1959
(millions of therms)

| | Natural Gas | | Manufactured Gas | | Percent of |
Year	U.S.	Northeast	U.S.	Northeast	Total[1]
1935	10,635	1,155	1,611	969	60
1937	13,480	1,478	1,535	999	65
1939	13,576	1,310	1,580	1,058	67
1941	16,358	1,469	1,726	1,145	66
1943	20,325	1,766	1,967	1,308	66
1945	22,563	1,647	2,088	1,382	66
1947	26,022	1,974	2,319	1,617	70
1949	32,234	2,219	2,274	1,696	75
1951	44,718	3,304	1,763	1,374	78
1953	52,800	4,273	838	608	73
1955	63,008	5,504	457	283	62
1957	74,649	6,887	215	120	56
1959	85,518	8,495	143	76	53

NOTES: Northeast includes Connecticut, Maine, Massachusetts, New Hampshire, Rhode Island, Vermont, Pennsylvania, New York, and New Jersey.

A therm is equivalent to 100,000 Btu.

[1]Percent of Total is the percentage of manufactured-gas sales in the Northeast compared to total manufactured-gas sales in the United States.

SOURCE: American Gas Association, *Historical Statistics of the Gas Industry* (Arlington, VA: AGA, 1964).

pipelines, the Big Inch and Little Big Inch, which extended from Texas to New York. Texas Eastern won the bid, converted the lines to natural gas transmission, and began efforts to expand the peacetime gas industry to Philadelphia. Pennsylvania's strongly pro-coal government impeded Texas Eastern's plans until company officials personally wrote and promoted a federal eminent domain bill for natural gas pipelines.

Transcontinental, an unsuccessful bidder for the Inch Lines, targeted a new pipeline for New York City, the largest manufactured-gas market in the nation. In New England, Texas Eastern and Tennessee Gas battled each other in the regulatory arena, the courts, and the media to contract for gas sales in the metropolitan areas. For natural gas to enter these markets, the local utilities had to forgo their century-old manufactured-gas plants in favor of southwestern natural gas; both fuel efficiency and local regulatory pressure led these utilities to convert to natural gas.

The contest for the northeastern gas market was not a simple free market play. The competitive process reflected decisions made in the hearing rooms and back rooms of federal, state, and local regulatory agencies. Without a certificate of public convenience and necessity, no pipeline could be constructed or operated; the competition for northeastern markets centered on applications and hearings for these certificates. The certificate process exemplified both the strengths and the weaknesses of American-style democracy. All voices could be heard, any interested party could intervene, and lobbying was an accepted part of the overall process. Typically, competing pipeline companies, as well as coal and railroad interests, opposed every certificate application for an expanding pipeline system. The resulting hearings, which often lasted for months and occasionally years, shaped developments during the formative era in this highly capital-intensive industry.

THEORIES OF GAS PIPELINE INDUSTRY REGULATION

Numerous scholars have presented theories regarding industry regulation, and it is useful to review the basic concepts and their application in comparable industries before examining business and government relations in the natural gas industry. Thomas McCraw, in "Regulation in America: A Review Article" (1975), identified and surveyed two basic concepts of regulation: "public interest" and "capture."[10] Generally, public interest signifies government intention to remedy perceived defects in market operation through regulation. Conversely, capture, or private interest, models designate the regulated industry as the prime mover behind its own regulation; the industry desires regulation in order to perpetuate monopoly power or another market advantage. Pluralism offers a useful third approach by examining the interaction of various governmental and business groups in economic development.

The argument between proponents of public interest and those of private interest provides two distinct choices. Did the FPC between 1938 and 1954 successfully and for the benefit of society manage the rapid growth of pipelines from the Southwest to the Northeast? Or did pipeline companies control the regulatory process for their own benefit? In this case, the regulatory process does not fall neatly into either category. The struggle among various interest groups, private, public, or otherwise, suggests that a modified version of the pluralist model offers a more appropriate approach to understanding the regulatory process.

This work contends that the FPC's public hearings for granting a certificate provided a forum for an often complex interaction among various interest groups each attempting to define the market. The issue here is not the certificate itself. As Ralph K. Huitt noted, "The issuance of such certificates is a conventional function of public-utility regulatory bodies, and the Commission [FPC] has administered it in a conventional manner."[11] The significant issues are the certification process, the actual operation of business under certificate-style regulation, and the regulatory agency's success in allowing the market to function competitively while also controlling entry into the industry.

Existing literature tends to denigrate the ability of the certificate of public convenience and necessity to provide either an adequate procedural test for entry or a competitive business environment.[12] "These regulations," wrote Joseph P. Kalt, "have proven to be anticompetitive barriers to entry in numerous transportation sectors, including the natural gas, airline, and surface freight industries."[13] James R. Nelson's study "Entry into Power Markets" interestingly broaches the period 1930 to 1959 in the natural gas industry for a possible test case of his "entry" hypothesis, that competition and regulation may be made compatible. But Nelson noted that "the natural gas case has never provided a clear-cut parallel to the situation being analyzed here."[14] In *The Economics of Regulation: Principles and Institutions*, Alfred Kahn also indicated that there was little if any competition in the natural gas transmission industry in the postwar period. Kahn wrote: "For at least a decade after World War II, the energies of the industry and of the Federal Power Commission were directed principally toward extending supply to areas of the country not as yet fully served [with close to a nation-wide network]. The opportunities for growth in noncompetitive directions have declined and the possibilities of competition expanded correspondingly."[15] The noncompetition theory, however, is incorrect; intense competition for both certificates and markets clearly existed during the earlier formative period.

Certainly, interstate natural gas pipelines have near monopolistic powers, but in practice these pipelines tend to exist as natural oligopolies. Local gas distribution companies do have monopoly power in nearly every instance, but more than one natural gas pipeline typically serve those distribution companies. According to Breyer and MacAvoy, "The interstate pipelines have some of the characteristics of natural oligopolies—economies of scale in transmission would seem to justify no more than two or three pipeline sources of supply in any regional market with population less than 10 million."[16] Thus, competition between two or three pipelines seeking to serve a large market should be expected to occur, and did occur during the period in question.

INTERSTATE COMMERCE REGULATION: A BRIEF SURVEY

Prior to 1960, regulation in the United States consisted of two basic forms: (1) rate or price regulation and (2) entry regulation. After 1960 a slew of new regulatory authorities encompassing safety, consumer, and environmental concerns greatly broadened regulatory practice.[17] The first federal control over interstate commerce involved rate regulation of the railroads beginning in 1877. Rate regulation is the basic form of regulation, whereby the commission sets either minimum and/or maximum rates or prices for services. Federal entry regulation in interstate commerce came into being during 1920 and was commonly achieved through the regulatory agency's power to issue, or not issue, a certificate of public convenience and necessity.[18]

The application of rate and entry regulation to various industrial forms of interstate commerce represents a common strategy among various regulatory agencies. This view, however, differs with Thomas McCraw's statement that "more than any other single factor, this underlying [economic] structure of the particular industry being regulated has defined the context in which regulatory agencies have operated."[19] Instead, the political and economic character of interstate commerce appears to define the context of regulatory operation. Regulatory policy over interstate commerce is often shaped through a pluralistic process, when hearings are held, and revolves around two basic market issues: entry and rates.

Rate and entry regulation provided for in the certificate are common features applied to several interstate businesses by the Interstate Commerce Commission (ICC), Federal Power Commission (FPC), and Civil Aeronautics Board (CAB). To understand the common features of the certificate, it is useful to compare briefly the regulation of interstate industries. For the most part, the FPC and ICC regulated the bulk of the large interstate commercial industries. Besides natural gas, the FPC regulated the electric power transmission industry. The ICC regulated railroads, petroleum pipelines, and the trucking industry; the CAB oversaw the airline industry. Although each industry is different in both temporal development and economic structure, regulatory agencies imposed similar rate and entry guidelines on each.

The regulation of interstate commerce in the United States began with the Interstate Commerce Act of 1887. This act empowered the ICC to set just and reasonable rates for all interstate common carriers. The act applied specifically to common carrier railroads, the nation's largest industry, and prohibited a variety of discriminatory pricing practices which, wrote business historian Alfred Chandler, resulted in the creation of the ICC. Chandler acknowledged that "the regulatory commission has become as significant an institution in the American economy as the investment banking house, the large corporation, and the trade union."[20] Although the ICC's regulatory authority remained weak during most of its first nineteen years, the ICC provided the first model of a federal agency to regulate interstate commerce.

Under the Hepburn Act of 1906, oil pipelines became the next major industry subjected to ICC regulation.[21] The Hepburn Act prescribed that interstate oil lines would have the status of common carriers. Although the Hepburn Act applied to persons or corporations "engaged in the transportation of oil or other commodity . . . by means of pipelines," it specifically excluded the pipeline transmission of water and natural or artificial gas from ICC jurisdiction.[22] The ICC's impact on the petroleum pipeline industry between 1915 and 1931, however, was minimal. In fact, oil pipeline historian Arthur Johnson noted that the ICC did not decide a single legitimate petroleum pipeline rate case until 1944.[23]

The ICC acquired the power to regulate entry into the railroad industry for the first time under the authority of the Transportation Act of 1920. This act remanded railroads to private control after the government took control of the industry in 1917. The ICC's authority to regulate entry into the railroad industry was effected through the issuance of a certificate of public convenience and necessity. The certificate was required of common carriers for all extensions and new railroad lines.[24] During the remainder of the 1920s, no other significant legislative acts imposed regulation on any interstate industries. The railroad industry, however, began flexing its considerable political and economic muscle to encourage regulation of the trucking industry.

Mirroring the coming railroad opposition to the expanding natural gas pipeline industry, railroad companies, specifically the National Association of Railroad and Utility Commissioners (NARUC) of the 1920s, believed that the expanding and competitive trucking industry should be regulated.[25] Conversely, the trucking firms, except for a few of the larger companies, and the ICC did not agree. Although trucking escaped federal controls during the 1920s, the ICC determined during the depression that the public interest did require its regulation due to "the newness of the industry, ease of entry, absence of regulation, and insufficient operator awareness of costs [all of which] condemned the industry to chronic instability and excessive competition."[26] According to Alfred Kahn, the traditional belief was "that unregulated entry and price competition had resulted in poor service."[27] The growing trucking business threatened railroads; truckers enjoyed "geographic flexibility" and "route and rate flexibility arising from the absence of federal regulation."[28] Thus, the railroad industry lobbied for a comparable system of federal regulation of trucking in order to protect it "against the intensified competition of motor carriers and [to protect] motor carriers from one another."[29]

In response, Congress passed the Motor Carrier Act of 1935, which added trucking to the list of the ICC's regulated businesses.[30] In 1935, 28,000 existing common carriers applied for and received authority under a grandfather clause effective June 1, 1935, to continue operating under ICC jurisdiction; nearly fifty years later 15,000 carriers remained in operation, most of which were descended from the original 28,000. It was difficult for new carriers to enter the industry because they would have to prove that "the area cannot be served as well by existing carriers."[31] The act required common carriers to use "just and reasonable" rates and obtain a certificate of convenience and necessity. In order to acquire a certificate, the applicant had to prove it was "fit, willing, and able" (FWA) to operate and to prove that its service would satisfy public need.[32] According to Joseph Kalt, an applicant's demonstration of being fit, willing, and able "generally entails evidence of sufficient capitalization and insurance."[33]

The federal regulation of gas and electric power transmission also resulted from New Deal legislation. In response to the widespread corruption and inefficiency of the public utility industry, Congress directed the Federal Trade Commission (FTC) in 1928 to investigate it. The result was several pieces of legislation devised to deal with these problems. The Public Utility Holding Company Act (1935) generally imposed strict control on corporate structure and organization to prevent securities and accounting fraud. The Federal Power Act (1935) vested the FPC with the power to regulate the interstate transmission of electric power. The FPC's primary responsibility here was to oversee the voluntary interconnection and coordination of electric transmission systems and to ensure that rates charges were "just and reasonable."[34] In 1938, the

Natural Gas Act became law. It empowered the FPC to regulate the natural gas pipeline industry through certificates of public convenience and necessity. Producers and end-users did not fall under the act although the Supreme Court ruled sixteen years later that producers were subject to FPC regulation.

The Civil Aeronautics Act of 1938 created the Civil Aeronautics Board and provided for the regulation of the airline industry.[35] Originally, the CAB's primary function was the administration of the U.S. Postal Service Administration's subsidy of the industry; rate regulation was secondary. Rate regulation became substantially more important in the 1950s. The CAB also practiced classical regulation theory in that it issued certificates of public convenience and necessity to applicants who proved "fit, willing, and able" to perform the service required by the public interest and charged just and reasonable rates.[36]

According to Breyer, the one basic difference between the ICC and CAB was in entry policy. "For many years," Breyer wrote, "it was comparatively easy for an airline already in the industry to obtain permission to serve new routes. It was not easy for a trucking firm engaged in LTL [less than truckload] carriage to obtain permission to enter new territories." Indeed, the CAB would later prove to be even less tolerant of applications for new entry than the ICC. Between 1950 and 1974, the CAB did not grant a certificate to any of seventy-nine applicants for domestic scheduled air service.[37]

The Transportation Act of 1940 added domestic water carriers to the list of industries under ICC jurisdiction.[38] Again, while the ICC regulated the water transportation industry, it required that common carriers acquire from the agency a certificate of public convenience and necessity before engaging in water-going transportation. Common carriers in existence prior to June 1, 1940, could continue operations under the act's grandfather clause.

Although interstate natural gas and petroleum pipelines, trucks, railroads, ships, and aircraft all fell under different modes of regulation, similar rate and entry rules were applied through the certification process to each of them. As noted, entry restrictions are typically blamed for limiting competition and allowing higher prices while price regulation is commonly cited as a leading cause of artificial market dysfunction. Certainly, many supporting cases can be found in the history of every regulated interstate industry. It is the purpose of this study, however, to provide at least one example in which entry restrictions did not diminish competition or overtly distort market mechanisms. The interplay among the various interest groups, business concerns, and governmental agencies involved in the regulatory process suggests the possibility for successful pluralism. The standards set by regulatory tools such as the certificate of public convenience and necessity allow a forum in which different political and industrial forces have the opportunity to interact and produce a successful economic development based on market forces.

2. Prelude to Expansion

NINETEEN THIRTY-EIGHT was a pivotal year in the development of the natural gas industry. The newly empowered FPC prepared to regulate the interstate transmission and sale of natural gas as industry observers pondered when and how long-distance pipelines would connect massive southwestern gas reserves with large northeastern energy markets.[1] Among other contemporary publications, Louis Stotz and Alexander Jamison's collaborative book, *History of the Gas Industry*, published in 1938, recognized the possibility and importance of gas pipeline connections between the Southwest and Northeast. As part of a discussion on the growing effort to convert the preponderance of nineteenth-century manufactured-coal-gas utilities in the Northeast to natural gas, the authors wrote: "To what extent such a changeover may take place in the future, particularly in the thickly populated areas in the New England and Middle Atlantic states, will depend largely upon economic conditions and necessity; [and] the ability to finance and construct pipe line systems to transport natural gas over such long distances."[2]

By the late 1930s, the Northeast—eastern Pennsylvania, New York, New Jersey, and New England—was the last substantial energy-consuming region in the United States without access to natural gas. The Pacific Northwest and southeastern states had minimal access to significant quantities of natural gas, but those regions did not represent a demand for the fuel comparable to that of the heavily populated and industrialized Northeast. Gas pipelines then extended from the Texas Panhandle into the Midwest, along the Gulf Coast, in California, and throughout Appalachia.

Generally, inadequate pipeline technology limited the ability of the early natural gas industry to expand rapidly. Nineteenth-century pipelines made of wood or poor quality iron and connected with leaky coupling joints were incapable of transporting natural gas over long distances. As technology improved and financing became available, gas companies built longer pipelines. The case of Southwest to Northeast expansion, however, is a unique industrial episode in part because it involves the demise of the nineteenth-century manufactured-gas business as well as dramatic growth and competition in the regulated natural gas industry.

MANUFACTURED GAS

The United States gas industry actually first developed in the Northeast. This early gas business involved the production by gas works companies of manufactured gas. The gas works produced this primitive synthetic gas through a process first developed in Europe for producing flammable gases from coal.[3] Drawing upon European scientific and commercial success with the utilization of artificially produced gas, the portrait painter Rembrandt Peale and Benjamin Kugler formed the Gas Light Company of Baltimore on June 13, 1816, the first gas light company in America.[4] Their company's success soon led to the creation of other gas light companies.

By the early 1820s, entrepreneurs formed manufactured-gas works in many large northeastern cities. These consisted of a gas-manufacturing complex and a small pipeline distribution system used to transport the gas short distances from the local gas plant to the consumer. The earliest companies included the New York Gas Light Company (1825), the Boston Gas Light Company (1822), and several companies in Philadelphia (1836). Later in the century, manufactured-gas companies appeared in midwestern cities. By 1900 virtually every major American city had its own manufactured-gas works with the large northeastern cities remaining as the largest manufactured-gas-consuming region in the United States (see table 2.1).[5]

Gas plant operators used several different methods to make their product. The Baltimore company began producing gas by distilling pine tar. The resulting gas flowed from the retort, or oven, to a gasometer, a tank made from wooden planks and held together by iron bands inverted in water. By 1822, the standard English method of manufacturing gas from coal replaced the pine tar method. In the English process, an operator heated coal in an oven and captured the resulting distilled gas as well as other by-products such as coal tar and coke, which could also be

Table 2.1. Introduction of Manufactured Gas to American Cities

City	Year	City	Year
Baltimore	1816	Indianapolis	1852
New York	1823	Norfolk	1852
Boston	1829	Memphis	1852
New Orleans	1832	Milwaukee	1853
Louisville	1832	Atlanta	1854
Pittsburgh	1836	Toledo	1854
Philadelphia	1836	San Francisco	1854
Cincinnati	1840	Scranton	1857
St. Louis	1846	St. Paul	1857
Providence	1848	Portland	1860
New Haven	1848	Kansas City	1867
Buffalo	1848	Oakland	1867
Rochester	1848	Los Angeles	1867
Washington, D.C.	1848	Omaha	1868
Cleveland	1849	Minneapolis	1871
Detroit	1849	Seattle	1873
Chicago	1850	Tacoma	1885
Columbus	1850	Spokane	1887

SOURCE: Arlon R. Tussing and Connie C. Barlow, *The Natural Gas Industry: Evolution, Structure, and Economics* (Cambridge, MA: Ballinger Publishing, 1984), 13.

sold.[6] In the early 1870s, Professor Thaddeus S. C. Lowe of Norristown, Pennsylvania, developed a new process for manufacturing gas.[7] Lowe mixed the coal gas with a spray of oil to produce a 500 to 600 Btu gas.[8] Gas manufactured by this process generally delivered a relatively constant level of heating quality to customers.[9] Despite these numerous improvements, however, the basic end-product remained essentially the same through the early 1950s.[10]

NATURAL GAS

The early American natural gas industry developed along a very different path. Natural gas had been used commercially in the United States as early as 1821 in Fredonia, New York, but natural gas consumption did not become widespread until after 1859. In that year, Colonel Edwin L. Drake discovered large quantities of both oil and gas in Pennsylvania.[11] Although oil, not natural gas, attracted the most interest from entrepreneurs, the latter product still found markets.

After an oil discovery, well operators typically allowed associated natural gas to escape into the atmosphere as mere "waste gas"; its only practical value was to pressurize the well and force oil to the surface. The high energy value of natural gas — approximately 1,020 Btu — eventually attracted local interest, and gas fields increasingly became an important source of energy for local industry. Beginning in the 1860s and lasting through the second decade of the twentieth century, Appalachian area gas reserves were the largest in the United States.[12]

The primary problem confronting the early natural gas industry was primitive pipeline technology. Customers could not purchase natural gas unless a voluminous field existed within a fairly short distance from the city; poor pipeline technology prohibited the construction of long-distance leakproof lines. Several Appalachian gas companies did organize pipelines to deliver local natural gas production to nearby cities, but many of these early efforts ended in failure due primarily to leaky pipelines.[13] Gradually, gas companies built longer lines, but before 1925 the longest one was less than 200 miles.[14]

A second difficulty confronting the industry was the accurate determination of a gas field's reserves. Often, a particular gas well would be depleted within only a few years after its discovery, causing reliability problems for consumers. In Indiana, these problems became especially severe. During the autumn of 1886, natural gas quickly began to replace the manufactured variety used for lighting state-wide. By 1907, however, many Indiana gas fields were practically depleted, forcing the local gas companies to return to manufactured gas. Without reliable gas reserve estimation techniques, similar problems occurred regularly and represented an unreliable element of the emerging natural gas industry.[15] The recurring fears of shortages combined with unreliable field reserve estimation techniques inhibited the early growth of the natural gas industry.

During the early 1900s, the changing geographical location of natural gas reserves and developments in long-distance pipeline technology dramatically altered the market structure of the natural gas industry. In Appalachia, gas reserves leveled off as new discoveries only replaced current production (see table 2.2). At the same time, gas and oil exploration in the Southwest resulted in increasingly larger gas finds. Discoveries of natural gas in the Southwest during the early 1900s led to the establishment of enterprises such as the Lone Star Gas Company, Kansas Natural Gas Company, Cities Service Gas Company, and United Gas

Table 2.2. Natural Gas Production by Region, 1912–1970

| Year | Percentage distribution by region | | | Total Marketed Production (tcf) |
	Appalachia	Southwest	Other	
1912	74	22	2	0.56
1920	55	34	11	0.80
1922	46	37	17	0.76
1924	31	45	24	1.14
1926	26	50	24	1.31
1928	21	57	22	1.57
1930	17	61	22	1.94
1935	16	65	19	1.92
1940	15	68	17	2.66
1945	10	73	17	3.91
1950	6	80	14	6.28
1960	3	87	10	12.80
1970	2	90	8	21.90

NOTES: Appalachia includes Pennsylvania, Ohio, West Virginia, and Kentucky (and New York for 1920 only). Southwest includes Texas, Louisiana, Oklahoma, and Kansas.

tcf is equivalent to trillion cubic feet.

SOURCES: U.S. Bureau of Mines, *Minerals Yearbook* (Washington, DC: GPO), and David Gilmer, "The History of Natural Gas Pipelines in the Southwest," *Texas Business Review* (May–June 1981), 133.

Company, which would later evolve into some of the nation's largest gas companies.

During 1918 and 1919 drillers discovered two huge southwestern natural gas fields: the Panhandle field in north Texas and the Hugoton field in the mutual border areas of Kansas, Oklahoma, and Texas. The combined Panhandle/Hugoton field would soon become the nation's largest gas-producing reserve, comprising more than 1.6 million acres. It contained original reserves of 117 trillion cubic feet (tcf), which accounted for approximately 16 percent of total U.S. reserves.[16] But as oil men had done earlier in Appalachia, they initially exploited the Panhandle field for petroleum only and allowed more than 1 billion cubic feet per day (bcf/d) of natural gas to escape into the atmosphere. As in Appalachia sixty years earlier, the commercial value of natural gas soon attracted entrepreneurial interest.

These discoveries encouraged both increased market demand and advancements in pipeline technology. In the early 1920s, pipeline builders began using acetylene torches to weld pipelines, and in 1928 electric welding came into widespread use. Electric welding in particular allowed for mass production of thin-walled, large-diameter pipelines of high strength which were both practical and economical. Generally superior welding techniques as well as improved pipeline sealing made pipe joints stronger than the pipe itself.[17] These newer, strong-walled lines were leakproof at very high pressures, allowing them to transport for longer distances much higher volumes of compressed natural gas. Concurrent with improvements in pipeline construction materials and techniques, gas compressor and ditching machine technology improved. The modern gas pipeline industry emerged during the 1920s.[18]

Discoveries of massive southwestern gas reserves and development of pipeline technology threatened Appalachia's position as the primary gas-producing region. As late as the 1920s, virtually all interstate transportation of natural gas still took place in the heavily populated Northeast, which relied upon Appalachian production. As one scholar of the industry noted, of the 150 billion cubic feet of gas moved interstate in 1921, about 65 percent was produced in West Virginia. Most of the gas flowed into Pennsylvania and Ohio. Less than 2 percent of the total interstate movement of gas originated in Texas.[19] With the gradual drying up of Appalachian fields beginning in the 1920s, state legislators attempted to prohibit the exportation of gas from one state to another in order to protect each state's industry. In West Virginia, the state congress designed the Steptoe Act to block out-of-state sales, and Oklahoma proposed a similar act.[20] These attempts to block out-of-state sales were largely unsuccessful.

RAPID EXPANSION IN THE GAS PIPELINE INDUSTRY

Between the mid-1920s and the mid-1930s, the combination of improved pipeline technology, abundant and inexpensive southwestern gas reserves, and growing nationwide demand resulted in an explosion of long-distance pipeline construction emanating from the Southwest (see map). Metropolitan gas distribution companies, which were typically part of large holding companies, financed most of the pipelines built during this era. In 1927, Cities Service constructed the first large pipeline system originating from the Panhandle field. This was a 20-inch pipeline, 250 miles in length, connecting the Panhandle field with Wichita, Kansas. It had a delivery capacity of 70 million cubic feet per day (mmcf/d) with compressor stations located at 60-mile intervals on the

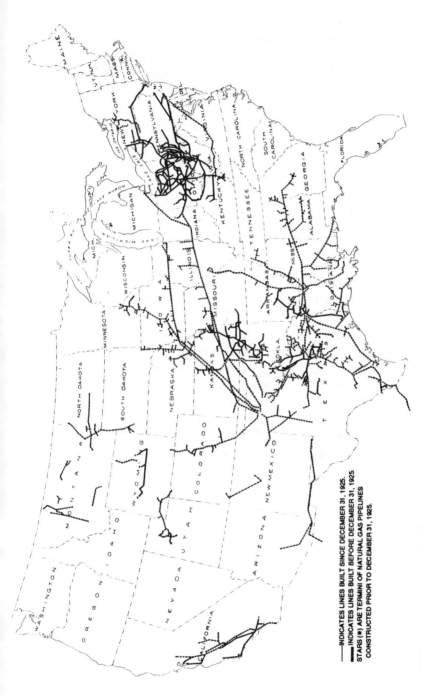

Principal natural gas pipeline systems in the United States, 1940. Source: Federal Trade Commission Monograph no. 36 on Natural Gas Pipelines in USA. Published as Temporary National Economic Committee Report no. 76-3 (Senate Committee Print). Washington, DC: U.S. Government Printing Office, 1940.

line. By 1928, the line extended directly into Kansas City.[21] In 1927, the Colorado Interstate Gas Company organized a 350-mile, 22-inch line originating in the Panhandle field which supplied cities in Colorado and Wyoming. Several other pipelines originating in Texas, Louisiana, and Oklahoma fields were also organized during the mid-1920s.[22]

All the most spectacular pipelines constructed during this period were in the 1,000-mile range. These long-distance lines generally originated in the Texas Panhandle and competed for midwestern markets. An era similar in many respects to that of the 1940s and early 1950s — except for the lack of federal regulation — the late 1920s pipeline boom caught the attention of government investigators of the industry. These observers described the race for midwestern markets as "an amazing story of high finance, suppression of competition, division of territory, and capture of control or forced receivership by established interests of independent enterprises which aspired to a share of the apparently large and profitable markets of the Middle West."[23]

Many of these lines had been designed during the 1920s, and all were completed during the Great Depression. In 1930, Samuel Insull, the captain of the American public utility industry, and others developed plans to build the Natural Gas Pipeline Company of America, to extend 980 miles from north Texas to Chicago. Insull's interest in the pipeline was to supply gas to his Chicago-area distribution systems. The first length was laid on August 22, 1930, and the last on August 5, 1931.[24] A second long-distance system emerging from the Panhandle area was the Panhandle Eastern Pipe Line Company. In 1930, it began construction of a gathering system and a 22-inch transmission line extending from Oklahoma into Indiana. Panhandle Eastern's parent company, the Missouri-Kansas Pipe Line Company, financed the 900-mile line from Texas to Indianapolis, which was supplied by the Hugoton and Panhandle fields. The line went into service in the early part of 1932 and was first to connect Texas gas directly with Appalachian customers.[25] The Northern Natural Gas Company was the third long-distance pipeline built in the late 1920s. It was a 1,110-mile line which connected the Panhandle/Hugoton field with Minneapolis via Omaha and served various other cities along its path.[26]

As new long-distance pipelines reached outward from the Southwest to the Midwest, well-established Appalachian-area gas distribution systems continued to sell regionally produced natural gas to local customers. Numerous small gas companies had grown in Pennsylvania after the 1859

Table 2.3. Average Retail Prices to Domestic and Commercial Consumers of
Competitive Fuels in the United States, 1935–1947
(per million Btu)

Year	Natural Gas	Manufactured Gas	Bituminous	Anthracite
1930	$0.679	$2.158	$0.341	$0.538
1935	0.669	2.052	0.320	0.439
1940	0.657	1.991	0.326	0.440
1941	0.628	1.848	0.345	0.471
1942	0.610	1.806	0.360	0.491
1943	0.594	1.750	0.377	0.520
1944	0.584	1.742	0.389	0.549
1945	0.601	1.696	0.397	0.568
1946	0.595	1.710	0.415	0.635
1947	0.594	1.698	0.549	0.709

SOURCE: Federal Power Commission, *Natural Gas Investigation, Docket No.
G-580*, Smith & Wimberly Report (Washington, DC: GPO, 1948), 338–39.

oil and gas discoveries, and such companies often merged to form larger
systems. By the early twentieth century, two large Appalachian distribution systems dominated the regional gas industry. These were the Standard Oil Company of New Jersey (later known as Consolidated Natural
Gas) and the Columbia Gas & Electric System.[27] These two companies
supplied natural gas to customers throughout Appalachia.

A few major midwestern cities including Chicago, Minneapolis, and
Indianapolis received natural gas through interstate pipelines, but the
manufactured-gas industry retained its dominance in the Northeast.[28]
The natural gas industry did not expand during the 1930s, but the existing
pipelines made their fuel available to the many cities that previously did
not have it. Less expensive and more efficient, natural gas quickly began
to displace manufactured gas in those cities that had access to both fuels
(see tables 2.3 and 2.4). By 1932, more than 80 percent of nationwide gas
sales by volume consisted of natural gas compared to 20 percent for
manufactured gas. In some instances, utilities with new access to natural
gas gradually converted their systems from the manufactured to the
natural variety by producing mixed gas, a combination of both which had
a higher Btu content than manufactured gas alone.

During the 1930s, Chicago became the largest city to convert its utility
distribution system to mixed gas, and this effort foreshadowed larger-
scale conversions that would take place in northeastern cities in the next

Table 2.4. Major Cities Converting to Natural Gas by Year

City	Year
Los Angeles, CA	1927
Denver, CO	1928
Independence, MO	1929
Memphis, TN	1929
Muncie, IN	1929
Richmond, KY	1929
Salt Lake City, UT	1929
San Francisco, CA	1929
Albuquerque, NM	1930
Atlanta, GA	1930
Biloxi, MS	1930
Birmingham, AL	1930
Lincoln, NE	1930
Mobile, AL	1930
Montgomery, AL	1930
Springfield, MO	1930
Butte, MT	1931
Phoenix, AZ	1931
Omaha, NE	1932
San Diego, CA	1932
Sioux City, IA	1932
Buffalo, NY	1933
Casper, WY	1933
Pittsburgh, PA	1933
St. Paul, MN	1933
Tucson, AZ	1934
Des Moines, IA	1935
Detroit, MI	1936
Grand Rapids, MI	1936

NOTE: Date given is earliest conversion by one city utility.

SOURCE: Louis Stotz and Alexander Jamison, *History of the Natural Gas Industry* (New York: Stettiner Brothers, 1938), 299–302.

two decades. In 1931, the recently constructed Natural Gas Pipeline Company of America system (NGPL) connected Chicago's metropolitan gas distributor, Peoples Gas Light and Coke Company, a co-owner of NGPL, with southwestern gas. Peoples Gas Light began producing a mixed gas with a 800 Btu content.[29] In conjunction with the introduction of natural gas into its system, Peoples Gas Light organized the first significant attempt at mass merchandising gas house-heating equipment.

With a potential market of 1.25 million residences, the local gas industry engaged in an aggressive promotion of gas that characterized the fuel's increasing attractiveness to both industry and consumer. The utility placed full- and three-quarter-page advertisements in Chicago and fifty outlying newspapers, and paid for advertisements on billboards, street-cars, and in show windows. In addition, Peoples Gas Light hired 270 company-trained salesmen, 60 heating engineers, and 14 sales directors to promote gas use around the city. Within the first ten weeks of the promotion, the company made 10,000 installations consisting mostly of conversion burners, and the company made 30,000 gas installations during the promotion.[30] The utility had to adjust existing residential furnaces to accept the higher Btu mixed gas. In order to convert appliances, gas mains required cleaning to remove the oil residue and moisture from the manufactured gas, but conversion to natural gas led to reduced costs for both utility and customers.[31]

Regulation, for all practical purposes, did not exist for the interstate gas companies. While rapid expansion of the natural gas pipeline industry during the 1920s and early 1930s created an extensive commercial interstate pipeline system, state commissions could attempt, with little positive results, to regulate the lines. A state could not regulate or even investigate a pipeline's operations outside its territorial jurisdiction, and the large interstate lines owned by huge holding companies were adept at disclaiming state regulatory power over their interstate operations. Consequently, "critical financial data and crucial cost information (e.g., the utility's power or gas purchases from sister companies outside the state) were not readily available to the [state] commission," rendering effective regulation "difficult, if not impossible."[32] Unbridled growth and corruption often resulted.

CORRUPTION, DEPRESSION, AND REGULATION

The depression halted this first dramatic phase of expansion in the natural gas industry. Other than the pipelines built during the late 1920s and early 1930s, no other long-distance lines were constructed until World War II. Between 1932 and 1936, companies built only three major pipelines, none of which was longer than 300 miles. Not a single significant pipeline appeared from 1937 through 1942. Although pipeline technology existed to connect distant supply and consumption regions, the continuing financial stagnation of the 1930s prevented major pipeline construction and, consequently, industry growth.

During the depression, many pipelines operated at less than 50 percent capacity; some companies had built pipelines during the 1920s and 1930s before securing markets, and others simply suffered from a diminished gas demand.[33] According to one scholar, the natural gas industry was in chaos during the depression: "In the East, it was marked by monopoly, shortage, and increasing prices. In the Southwest, there was an enormous oversupply," a great deal of which was allowed to dissipate into the atmosphere.[34]

In addition, the corporate and financial structuring of utilities under the holding company system revealed during the depression a wide array of corrupt accounting and financial practices. Generally, a single holding company would own several gas and electric companies, which in turn would control additional utility companies, and so on. Through various forms of stock fraud, the holding company would control a vast empire of utility companies while simultaneously raising a substantial amount of money from essentially worthless stock issues. These utility pyramid empires, such as Samuel Insull's, came crumbling down during the depression.[35]

These and other problems affecting the entire utility industry including natural gas evoked strong congressional response even before the stock market crash of 1929. The public utility industry had become so large, powerful, and dubiously successful that on February 15, 1928, the Senate directed the FTC to report on the condition of the existing public utility holding companies.[36] The FTC's investigation and its massive report, consisting of ninety-six volumes and printed in 1935, became the basis for a wide range of legislation directed at the regulation of the utility industry.[37]

The FTC report addressed problems confronting the entire utility industry. In regard to natural gas, the report clearly indicated that a lack of conservation, meaning waste or inefficient use, was the natural gas industry's most pressing problem and that conservation efforts might be regulated through gas well drilling and production controls. There was also discussion of making all gas pipelines common carriers, as the Hepburn Act did for oil pipelines.[38] The issue of conservation was particularly important because natural gas was recognized by the 1930s as an exhaustible resource. The FTC report found an astounding level of waste, which was no secret to those in the industry (see table 2.5). Practically all the waste was due to oil well operators allowing unwanted natural gas associated with the desired oil simply to escape into the atmosphere.[39]

Table 2.5. Estimated Waste of Natural Gas in the United States
(bcf)

Year	Total U.S. NG Waste	Panhandle Texas NG Waste	Total U.S. NG Consumption
1919	213	n/a	256
1920	238	n/a	286
1921	193	n/a	248
1922	233	n/a	254
1923	416	n/a	277
1924	343	n/a	285
1925	324	n/a	272
1926	417	220	289
1927	444	405	296
1928	412	351	321
1929	589	294	360
1930	553	252	376

SOURCE: Federal Trade Commission, *Report to the Senate on Public Utility Corporations*, Senate Document no. 92, 70th Cong., 1st sess., 1935, pt. 84-A, 93, 95.

In addition, the report identified sixteen "evils" existing in the natural gas industry. They included (1) the "great waste" of natural gas, (2) the unregulated structure of utility companies, (3) the pyramiding corporate structure of utility companies, and (4) excessive profits, inflation of assets, and stock watering. The FTC report stated that federal regulation of the industry was necessary because individual states could not effectively regulate interstate firms. According to the report, the states' ability to regulate natural gas was "at best indirect, partial, and poorly founded because of their limited authority to ascertain facts and their lack of authority to regulate interstate commerce."[40]

The inability of state commissions to regulate the interstate industry did not result in a workable free market system. Other than the evils of pyramiding, stock watering, overvaluation, and waste, the ability of any one company to compete for a new market depended upon what the FTC report described as a traditional "system of so-called 'ethics' which hold that it is unfair and unethical for a pipeline to invade any territory already being served or claimed by another line. Such invasions are classed by the industry as 'raids' and the fear of retaliation by powerful interests is usually sufficient to preclude them."[41]

Excessive monopoly power appeared to the FTC to be at the heart of these problems. The investigation revealed that four dominant holding companies controlled directly and indirectly more than 60 percent of the natural gas produced and 58 percent of the total pipeline mileage.[42] These companies also represented the changing locus of geographic market structure. The largest utility holding company was Columbia Gas & Electric, which operated gas transmission and distribution systems in the Appalachian region. Also in Appalachia, Standard Oil of New Jersey controlled the Consolidated Gas System. The southwestern region included the two other large holding companies: Cities Service, which operated in Kansas and the Panhandle area, and Electric Bond & Share's primary gas system, the United Gas Corporation, which operated primarily in the Texas and Louisiana region.[43]

In response to the FTC's investigation, Congress passed several legislative acts. These included the Public Utility Holding Company Act of 1935 (PUHCA), the Federal Power Act of 1935 (FPA), and the Natural Gas Act of 1938 (NGA). In addition, the Securities and Exchange Commission (SEC) was charged with protecting the utility security holder through various regulations.[44] Congress designed the PUHCA, FPA, and NGA to become law simultaneously. However, the original legislation affecting the natural gas industry was significantly modified before becoming law in 1938.

Generally, the PUHCA required all utility systems consisting of more than three tiers of companies to be abolished, and it limited individual companies to a single, integrated public utility system. The act also required all utilities to register with the Securities and Exchange Commission, provide it with detailed financial data, and receive its approval before undertaking most financial transactions, from issuing securities to reorganizing. Many large utilities ignored these rules until 1938, when the Supreme Court upheld the PUHCA.[45]

The foremost result of the PUHCA was that the large holding company systems divested many of their subsidiaries, including pipeline interests. The PUHCA prevented the large gas distribution companies from owning and controlling both pipeline and production companies. Pipeline companies acquired independence and competed against one another for survival without financial or managerial support from a parent company. Between 1935 and 1947, holding companies sold off 306 utility companies including 113 gas companies. Immediately after 1935, pipelines began seeking institutional investors such as banks and insurance companies to purchase their securities. Life insurance companies, in particular, fat with cash but lacking secure financial investment opportunities during the immediate postdepression years, needed good investments. Richard Hooley described the investment situation of depression-era insurance companies: "The shrinkage in the volume of top quality corporate bond issues coming to market together with the sharp decline in the yield on life insurance investments resulted in a situation where life companies experienced increasing difficulty in acquiring investments yielding returns sufficient to meet commitments on outstanding policies."[46] The newly independent and rapidly growing natural gas pipeline companies proved attractive to the insurance companies interested in new investment opportunities.

The Federal Power Act was the second piece of New Deal legislation affecting the utility industry. Drafted by the FPC under an executive order issued by President Roosevelt, it specifically addressed the unregulated interstate electric power industry.[47] The Federal Power Act generally imposed rate regulation on the interstate electric transmission industry and encouraged the interconnection of the industry. The FPC's new role seemed to be a natural fit. Congress originally created the Federal Power Commission under authority of the Water Power Act of 1920 to regulate the burgeoning water and hydroelectric power industries. The Water Power Act of 1920, under consideration in various forms for nearly ten years, was the result, according to O. C. Merrill, a strident supporter of the then proposed agency and chief engineer of the Forest Service, of abnormal industrial and financial conditions brought about by the war. Noting the surge in demand for electric and steam power during World War I, Merrill noted that "if the government takes no action, such legislation [the Water Power Act] is likely to bear the imprint of the active but unrepresentative minority of the water-power interests. If, on the other hand, the Government will take the initiative, it can, with the cooperation and support of a substantial majority of the water-power interests, draft a bill which will adequately protect both capital and the public."[48] Another influential supporter was Gifford Pinchot, who wrote to Congressman Scott Ferris: "I regard it as of very great importance, as you do, that the bill should not fail."[49]

The original FPC was composed of five directors, including the secretaries of war, interior, and agriculture. Its function was the licensing of all water power developments on national public land or on navigable waters subject to national jurisdiction. In 1930, the FPC became an independent body with five full-time commissioners. Although the Federal Power Act gave the FPC an important new mission to regulate the interstate electric power transmission rates, the FPC would soon acquire an even more important role.

The Natural Gas Act of 1938 was the most significant New Deal legislation directed at the natural gas industry. Congress designed it as a stabilizing force on the operations of newly independent pipeline companies.[50] The Natural Gas Act, however, followed by several years the FPC's acquisition of regulatory power over the interstate electric power industry.[51] After receiving the FTC report, President Roosevelt specifically requested the FPC to draft electric power legislation; neither the president nor Congress requested any specific natural gas legislation. But state regulatory commissions pushed federal regulation of natural gas, as exemplified by a letter to that effect sent to Congress by John E. Benton, general solicitor of the National Association of Railroad and Utility Commissioners (NARUC).[52] State regulatory commissions actively lobbied for federal natural gas legislation to be submitted at the same time as the holding company and electric power legislation of 1935.

Congressman Sam Rayburn (D) of Texas then began the legislative process by giving a House legislative draftsman a copy of the Federal Power bill with instructions to revise it to apply to natural gas. Senator Burton Wheeler (D-MT), a sponsor of the gas legislation, stated in Congress that "the purpose of the bill is to help the state commissions and the people of the country find out what is the cost of transporting natural gas to the large cities."[53] Rayburn introduced to the House the resulting draft natural gas legislation as Title III of the PUHCA, but Rayburn did not introduce Title III to the Senate. Instead, it went to committee separate from the PUHCA and FPA. Later, Rayburn and Congressman Clarence Lea of California called on the FPC to participate in the final drafting of the legislation, which became the Natural Gas Act of 1938. In addition, Lea sent copies of the legislation as it was being drafted to all state commissions and the general solicitor of the National Association of Railroad and Utility Commissioners. State commissioners supported federal regulation of the industry in order that companies operating in the states without any regulation might be thereby controlled.[54]

A wide variety of business groups eventually supported the Natural Gas Act. M. Elizabeth Sanders has shown that the "designers of the Natural Gas Act were able to put together a broad-based supportive coalition." Generally, consumers, local distribution companies (LDCs), and producers supported the Natural Gas Act's restrictions on price although only pipeline companies and coal companies tended to favor entry restrictions. Pipeline companies relished the idea of limited intra-fuel competition, and railroad companies perceived entry restrictions on gas pipelines as at least one regulation which might limit coal and gas competition.[55] Indeed, the coal industry supported the Natural Gas Act with the same motivation that the railroad industry supported the Motor Carrier Act of 1935—to diminish interindustry competition. The coal industry "of all the advocates of the Natural Gas Act of 1938," according to historian John Clark, "insisted upon the most rigid regulatory formula."[56]

The coal industry supported regulation as a means to raise the price of natural gas and lessen its competitive effects. Since World War I, coal consumption had steadily declined due in large part to the increasing demand for natural gas. At the congressional hearings on natural gas held during 1936, John Battle, president of the National Association of Bituminous Coal Organizations, testified that "natural gas has displaced millions of tons of bituminous coal throughout the nation. . . . It is our opinion that natural gas is being sold for industrial purposes at prices that are unreasonably low. . . . We feel that the competition is unfair." At the same time, gas industry representatives charged that coal and railroad interests and their political supporters—including the state of Pennsylvania—often refused to grant gas pipelines right-of-way into "coal territory."[57]

When Congress began considering legislation in 1935 to regulate the gas pipeline industry, pipeline companies generally opposed regulation although during the next two years they became more supportive. According to Elizabeth Sanders, there were four major changes in gas regulation policy between 1935 and 1937 that led to wider support for gas pipeline regulation. These included "(1) the elimination of the common carrier requirement; (2) the modification of the section on cost determination; (3) the change in certification from gas field to distributing market; and (4) the addition of an exemption for industrial sales from the rate-suspension provision of section 4."[58]

The original natural gas legislation was redrafted and submitted to Congress in 1936. After a redrafted version of the gas bill failed congressional passage due to "disinterest," it was submitted for a third time. After the addition of several amendments, it finally passed Congress without dissent in June 1938. The president signed it into law on June 21, 1938.[59]

The Natural Gas Act empowered the FPC to order natural gas companies to approve and/or set "just and reasonable rates," extend and abandon service, improve facilities, keep extensive documentation of operations and finances, ascertain costs for rate making, suspend rates, and otherwise dictate the operations of any interstate gas company.[60] To administer these powers, the act required the FPC to conduct hearings on a case-by-case basis in order to grant, or not grant, a certificate of public convenience and necessity to a natural gas company. This certificate was necessary before any natural gas company could operate in interstate commerce. With a certificate, a pipeline company held a franchise over a particular service area, although the FPC could certify more than one line to serve the same local or regional gas distribution company and thereby grant oligopolistic powers to two or three lines.

The certificate represented, at least in part, an attempt to solve historical problems of unreliable supply and operational inefficiency. In many cases, newly connected gas fields dried up within only a few years of discovery causing severe energy shortages in those communities dependent upon gas. To gain a certificate the pipeline had to have under contract an adequate supply of natural gas, usually twenty years', appropriate for its sales market. It also required the line to have adequate financing for construction or expansion and charge only "just and reasonable" rates. In essence, the certificate was the only independent confirmation that the pipeline appeared to have excellent prospects for long-lasting and reliable service to the consumer. Indeed, "certificates became a sine qua non for the sale of bonds financing new pipeline ventures."[61]

One feature of the certification process which would soon cause large problems for the FPC was section 7 (c), which stated: "No natural-gas company shall undertake the construction or extension of any facilities for the transportation of natural gas to a market in which natural gas is already being served by another natural-gas company . . . unless and until there shall first have been obtained from the Commission a certificate."[62] This phrase, according to Congressman Lea of California, a strident supporter of the gas bill, meant that "before you can have competition in the same territory, a permit must be secured from the [FPC]."[63] It would become increasingly clear over time, however, that the FPC was equally uncertain of its jurisdiction to certificate pipelines into areas in which natural gas was not already being served.

In return for compliance with the certification procedures, the FPC allowed pipelines to set gas sales rates that reflected the line's cost of operation, including gas costs, and "rate base," or depreciated assets used in the transmission of gas, plus a 5.7 to 6.5 percent profit on its sales.[64] The apparent guaranteed rate of return, although not reflective of a formula of great profit, later attracted the term *cash cow* to pipeline companies operating during times of abundant supply and demand. In addition, the FPC had authority to dictate the specific customers of a pipeline and the allocation of sales volume to those customers. But the Natural Gas Act did not grant eminent domain powers to natural gas pipelines. The railroad industry, which depended upon coal transportation for one-fifth of its total revenues, often opposed any attempt by a pipeline company to cross underneath the tracks. Lack of significant right-of-way posed a significant problem to many pipeline companies attempting to extend service and would later present a major obstacle to the expansion of the gas industry into and beyond Pennsylvania.[65]

The current set of five FPC commissioners was then responsible for regulating interstate natural gas commerce, and only three years earlier the FPC had acquired similar powers over the electric power industry. These new responsibilities did not result in any drastic changes, however, at the commission level. The seated members of the FPC in 1938 were Claude L. Draper, formerly of the ICC bar and staff of the NARUC; Basil Manley, past U.S. Senate staffer and official of the New York Power Authority; Clyde L. Seavey of the California Railroad Commission; and John W. Scott, a former U.S. assistant attorney general. Leland Olds, a New Deal utility economist with experience on presidential commissions and at the New York State Power Authority, was appointed in 1939 to fill a vacant seat. Once a rather innocuous agency, the FPC was now charged with what would become an increasingly important responsibility to regulate the nation's interstate gas and electric transmission.

One of the FPC's most pressing concerns was eliciting the cooperation of its new charges. Clyde L. Seavey, acting chairman of the Federal Power Commission, attempted to do this at a meeting sponsored by the American Gas Association. He urged all interstate pipeline companies to comply with FPC jurisdiction. Seavey noted that the FPC had recently mailed questionnaires to all known persons and companies engaged in the natural gas business to ascertain which ones were subject to FPC jurisdiction. Many of the largest companies complied by the deadline, but many others did not. Seavey described a similar situation in which many companies required to file rate schedules with the FPC had not yet done so, and he urged full cooperation with the FPC by the natural gas industry. In his closing remarks, he stated, "If the industry will fulfill its obligations under the Act in the same spirit, immeasurable benefits from the enactment of the Natural Gas Act will follow for the public and industry."[66] Clearly, Seavey presented the newborn FPC as an agency with the best interests of the natural gas industry at heart.

In the late 1930s, the structure of the natural gas industry was far different than it had been only ten years before. Many long-distance pipelines skirted the nation's mid-section, southwestern natural gas reserves promised an abundant fuel, and federal regulators idealistically forecast a promising future for the industry. But one highly industrialized and densely populated area of the nation, the Northeast, continued to depend upon a century-old manufactured-gas industry.

By the end of the 1930s, however, the pipeline industry was not growing. In some markets, natural gas sales were down. The country remained economically weak. Observers of industry did continue to speculate on the plausible scenario of the possibility of natural gas sales into the Philadelphia and New York areas. *Fortune* magazine, which devoted its August 1940 cover story to the natural gas industry, noted that "as for markets, not only is the richest one in the country, the Atlantic seaboard from Philadelphia north, still virgin (as in the state of Wisconsin), but natural gas does not begin to envelop the territory it serves or through which it passes." While the article remarked that the undeveloped market areas promised future growth for the industry, "a portentous group of natural gas enemies" was likely to oppose the expansion of the industry.[67] These enemies included the continuing wastefulness of the natural gas industry itself, the soft-coal interests, the railroads which transported coal, and the "conservative" manufactured-gas industry.

Contemporary estimates noted that the northeastern markets could account for a 25 percent increase in total U.S. natural gas sales, but no significant pipelines had been built into those areas. Columbia Gas & Electric owned a small line extending into Paterson, New Jersey, and Tappan, New York, as well as into the Philadelphia area. It also owned the 20-inch Atlantic Seaboard pipeline, which extended from Kentucky and West Virginia into Virginia, Maryland, and on to a single customer at Coatesville, Pennsylvania, west of Philadelphia, and to Washington, D.C. But the Atlantic Seaboard line was not used to capacity. Natural gas interests accused coal and railroad companies of plotting to keep natural gas out of the Philadelphia area by suggesting that Appalachian area gas reserves were not sufficient to supply the region with natural gas, an assertion which was probably accurate.[68]

Although the Panhandle Eastern Pipe Line Company did not extend into the Northeast, it was the only system that extended from the Southwest to the western edge of the Appalachian region. Even Panhandle's early history was tangled in allegations that coal and railroad industries diverted it from its original destination, the Omaha and Minneapolis metropolitan areas. After its construction, Columbia Gas bought a one-half interest in Panhandle Eastern, reportedly to "keep cheap Texas gas out of Columbia's territory" according to the Missouri-Kansas Pipe Line Company, Panhandle's other owner and founding company. Ultimately, Columbia was forced to divest its interest in Panhandle in response to antimonopoly sentiment. Then, Columbia publicly discussed the possibility of building its own line from the South into the Northeast, although it did not follow through with these plans.[69]

World War II ultimately overwhelmed domestic politics. The exigencies of war required much closer government scrutiny of industry and imposed a new set of rules on American business. Perhaps the forced wartime cooperation—through the effects of either patriotism or strict government control—lessened the impact of the new utility regulations placed on the industry. As the gas industry moved to support the war effort, many within it recognized a special opportunity to expand operations under emergency war conditions and then to be well positioned for continued growth and expansion after the war. Natural gas entrepreneurs were ready to ride their war emergency pipelines through the war and into a dramatic new era of industry growth.

3. Entrepreneurship and the Appalachian War Industry

WORLD WAR II stimulated renewed expansion of the natural gas industry. This phase of growth occurred under a variety of war emergency agencies which essentially controlled the industry. Through these agencies, the federal government financed war production plants and fuel transportation systems and strictly managed the nation's energy supply. Normal peacetime market mechanisms alone did not guide expansion. Rather, the vital needs of a wartime economy, interpreted by various federal agencies, spurred industry growth.

During the war, factories required record quantities of gas and its by-products for the manufacture of vital steel, aluminum, high-octane gasoline, synthetic rubber, chemicals, and explosives, and for industrial and domestic heat as well as power generation. Residential gas demand increased as well for heating the many newly constructed homes built to house the growing military and domestic war-related labor force.[1] The surge in wartime energy demand, combined with government interest in financing energy systems, stimulated entrepreneurial activity aimed at taking advantage of government support for pipelines to transport fuel to the vital war production and refining centers.

APPALACHIAN GAS SUPPLY AND FEDERAL WAR PLANNING

The Appalachian region quickly became the center of American war production. Cities such as Pittsburgh, Youngstown, and Wheeling contained hundreds of steel mills and metallurgical factories, as well as rubber and chemical plants which required large volumes of natural gas.[2]

Table 3.1. Estimated Gross Production of Natural Gas in Appalachia
(bcf)

Year	NY	PA	OH	WVa	Total
1937	22	120	46	163	351
1938	41	86	39	150	316
1939	31	104	44	172	351
1940	15	102	50	205	372
1941	13	105	51	223	392
1942	11	105	55	232	403
1943	9	103	63	243	418
1944	8	100	57	204	369
1945	10	90	55	182	337
1946	7	93	61	201	362

SOURCE: American Gas Association, *Historical Statistics of the Gas Industry* (Arlington, VA: AGA, 1964).

Natural gas was a particularly important fuel because it burned at a constant specific temperature, allowing for high-quality product manufacture. Approximately 660 factories in the Appalachian area required more than 24 bcf per year of natural gas. As a region, Appalachia required 400 bcf per year of gas, about half of which was accounted for by industrial consumption.[3] War hastened the decline of Appalachia's gas reserves; as supplies dwindled, new sources, especially those in the Southwest, became vital to the war effort.

The Appalachian region was the first in the United States to produce large quantities of natural gas, and its production peaked in 1917 at 552 bcf, or approximately 63 percent of the nation's total natural gas production. By the late 1930s, Appalachian production accounted for only about 16 percent of total U.S. gas production. This decline resulted from an overall decline in Appalachian gas reserves as well as a dramatic increase in southwestern-produced gas. Local gas companies continued to produce a significant quantity of natural gas for Appalachia, but by 1943 production could not keep up with the increasing industrial, commercial, and residential demand (see table 3.1).[4]

During the war, the Appalachian region consumed about 18 percent of the nation's natural gas production even though it contained only 3 percent of the nation's natural gas reserves. Additional supplies were needed to meet demand, but many Appalachian gas fields were rapidly depleting. In 1940, for example, gas companies drilled approximately 1,627 new wells in

Appalachia, but 25 percent of them were dry. The major prewar source of supply was the Oriskany Trend, a deep geologic formation which reached peak production in 1942 and began to decline rapidly after 1943.[5] Federal authorities initially attempted to increase local production, but later diverted the steel necessary to build pipelines to the war effort.

The intense drain on regional production soon stimulated private plans for both increased production and access to new supplies. As the industry expanded during the war, executives kept one eye on their industry's postwar future. This was particularly evident in speeches made at the American Gas Association's annual meetings. Gas company executives tended to be optimistic about the postwar demand for their fuel. At one meeting, J. French Robinson, a prominent gas utility executive, said, "In the post-war sunshine of abundant materials for our use, we will be able to realize the potential values of natural gas to all this nation as never before."[6] Natural gas consumption surged during the war years, especially because of high demand from war industries. But the gas industry as a whole did not seem worried that demand would fall off dramatically in the postwar period. The more industry, commercial establishments, and residential consumers used gas the more they would want to continue using it. The AGA also worked hard at encouraging and coordinating advertising efforts to further promote the use of natural gas.

As the European war intensified and American involvement became inevitable, direct government intervention in the nation's energy industry became unavoidable. During the early years of the European war, the American Gas Association and various other industry groups worked privately to promote growth, efficiency, and stability in the gas industry. Maintaining a high level of Appalachian natural gas production without incurring strict governmental control of the industry was an immensely difficult task.

On the government side, Secretary of the Interior Harold I. Ickes lobbied President Franklin D. Roosevelt to create an agency to oversee the nation's wartime energy industries. Confronting an unplanned domestic energy situation, Roosevelt chose Ickes to be the petroleum coordinator for national defense on May 28, 1941. This position gave Ickes special powers over virtually every aspect of the oil and gas industries. It also earned him the unofficial title of oil "czar" for his alleged socialistic ideas and preference for a nationalized oil industry.[7] Ultimately, Ickes carefully crafted a cooperative governmental business approach in his administration of the wartime industry.

Ickes's new organization, the Office of Petroleum Coordinator (OPC), developed later into the Petroleum Administration for War (PAW). The agency sought to elicit cooperation among oil and gas companies. To achieve this end while mollifying industry opposition to the agency, Ickes appointed Ralph K. Davies of Standard Oil of California as his deputy administrator. The PAW divided the United States into five districts, each overseen by a district chairman. Within this geographical framework, Ickes administered the ten separate divisions within PAW, as well as the activities of each district chairman.

The PAW put oil, and for good reason, higher than natural gas on its list of priority fuels. During the war, recalled J. R. Parten, director of transportation for the PAW, "Natural gas didn't stand very high, didn't take much of Ickes' time . . . natural gas was not a hot spot. Production of crude oil was."[8] Everette DeGolyer, PAW chief of petroleum conservation, was equally concerned about oil supplies. Bill Murray, conservation engineer for PAW District 3 (Texas, New Mexico, Louisiana, Arkansas, and Mississippi), recalled that DeGolyer's orders were simple: "Murray, you find out where it can be produced without waste. What's the maximum rate at which we can produce these fields?"[9]

Natural gas, though not as ultimately important as oil, was certainly a vital fuel. In one early meeting of Ickes's new agency, on December 4, 1941, only three days before Pearl Harbor, the participants met to discuss natural gas supply and demand in the Appalachian area. Representatives from a variety of agencies including the War Department, Navy Department, and FPC attended. Less than a few months earlier, the Office of Production Management (OPM, which later became the WPB) had begun receiving a slew of letters from Pittsburgh area industrial firms requesting natural gas. In one letter from the Aluminum Company of America (ALCOA), the firm asserted that it desperately needed natural gas in order to continue operations. "You understand," the company official wrote, "we are engaged in the production of defense work 100% and it is very essential that we are furnished gas to carry on our production."[10] Thus, conferees heard a supporting presentation: local Appalachian gas supply could meet regional war industry demand through the winter of 1941–42, but a new source of supply, such as would be afforded by a proposed pipeline connecting southwestern supply with Appalachia, would be necessary.[11]

After the OPC officially became the PAW, the PAW created a Natural Gas and Natural Gasoline Division to be responsible for overseeing developments in the gas industry. E. Holley Poe, a former executive of the American Gas Association and future gas industry entrepreneur, headed the division. His most important goal was maintaining production and deliverability of natural gas, particularly in the Appalachian region. Poe also attempted to marshal support for joint-industry cooperation while administering the wartime industry. At one American Gas Association meeting in 1942, Poe told the audience, "We are not approaching our job as a dubious experiment. . . . We do not like red tape any more than you do." Poe agreed with Ickes that the most satisfactory results for the wartime industry could be attained "by a mutual, unselfish understanding of the problems at hand and a cooperative spirit in their solution."[12] These statements reflected the government's efforts to control the energy industry through cooperative rather than confiscatory policies.[13]

The PAW's war emergency powers over natural gas were relatively modest compared to those of the Supply Priorities and Allocation Board (SPAB). The SPAB, which later merged into the War Production Board, had authority to control all areas of industrial production and supply. The agency dictated specific gas sales allocation orders to natural gas pipelines depending upon the fuel requirements of their war industry customers. During the latter part of 1941, representatives of the natural gas industry, military, PAW, WPB, and the American Gas Association met several times in different cities to discuss recommendations for limiting unnecessary natural gas consumption and maintaining natural gas production levels during war.[14] There was little argument about the possibility of serious gas shortages. J. A. Krug, chief of the power branch of the WPB, sent a memorandum to his superior, J. S. Knowlson, outlining a proposed "Limitation Order" to curtail the consumption of natural gas and thereby conserve it. Krug noted that serious shortages were likely, especially in Appalachia as well as in southern California and the mid-continent areas. The proposed limitation order, therefore, would have two major goals: increase gas production and curtail nonessential uses of gas.[15] A high-powered letter of support came from Major General H. K. Rutherford regarding the critical situation faced by war industries dependent upon natural gas.[16]

As a result of what appeared to be an impending fuel crisis, the WPB issued on February 16, 1942, its first order imposed on the natural gas industry, Order L-31. On February 25, the WPB mailed copies of the order to all utilities it affected. L-31 requested utilities to comply voluntarily with a provision calling for pooling arrangements "to achieve practicable maximum output in the area or areas in which a shortage exists or is imminent."[17] The order also provided the WPB with authority to integrate natural gas systems, curtail gas sales when necessary, and reallocate existing gas sales. Furthermore, the WPB actively encouraged pipelines to transport gas at 100 percent load factor, to use gas storage fields whenever possible in order to free up pipeline capacity for gas transmission, and to develop curtailment schedules that would affect the customers least dependent upon natural gas. In the summer of 1942, the WPB issued a similar order, L-174, modeled after L-31, which imposed the same restrictions on the manufactured-gas industry. Essentially, both L-31 and L-174 limited new gas sales for nonmilitary or nonmobilization purposes.[18]

The PAW and WPB also attacked the Appalachian gas production problem. Initially, the PAW issued conservation guidelines for new drilling programs in order to develop a nationwide oil and gas development drilling program "consistent with the availability of material and equipment."[19] The PAW's conservation Order M-68 implemented this program by limiting the drilling of gas wells to not more than one for every 640 acres. However, geologists determined that the maintenance of prewar levels of production required new drilling for shallow gas. Industry leaders expressed concern that Order M-68 would restrict new drilling and seriously threaten current production levels. In response, the PAW revised its spacing provisions on February 13, 1942, to permit the drilling of one well to each 160 acres for specified deep horizons and one to each 40 acres for shallow wells.

The importance of Appalachian natural gas supply to the war effort was reflected in the disproportionate number of gas wells drilled there. Between 1942 and 1945, approximately 70 percent of all gas wells drilled in the country were drilled in Appalachia. Increased drilling activity, however, did not significantly raise production levels. In lieu of increased reserves, federal restrictions placed on natural gas consumption sought to ensure the continued availability of natural gas. But such measures could not meet extremely heavy wartime energy demands on the Appalachian natural gas industry. Wartime demand aggravated a situation already characterized by rapidly diminishing reserves. Industries requiring increasing amounts of natural gas purchased more of the fuel than normal. Consequently, the Appalachian fields could not withstand for long the demand placed upon them for both wartime industry and existing residential and commercial utilization. Government drilling and consumption regulations did little to alter a dangerous energy shortage.

GAS PIPELINES AND APPALACHIAN MARKETS

The wartime shortage reinforced the belief held by some gas experts that the time was right to build a pipeline extending from the Southwest to Appalachia. Curtis B. Dall led one group in a significant attempt to build such a line. Dall, a former Wall Street broker with Lehman Brothers, was best known for his failed marriage to President Franklin D. Roosevelt's daughter, Anna. They had married during the late 1920s and were first estranged and then divorced by the mid-1930s. According to FDR biographer Kenneth Davis, Anna's "marriage to Curtis Dall had been a ghastly mistake; she could not bear to live with him."[20] Even after the divorce, Dall believed that business and government officials regarded him distrustfully due to his former relationship with Franklin Roosevelt, but he continued to pursue business opportunities.

Dall saw in the wartime gas industry a ripe entrepreneurial opportunity. After visiting Nashville, Tennessee, in connection with a business deal involving phosphate properties, Dall recalled that he "ran into a project which greatly interested me. Some friends described how nice it would be" to have natural gas in the Nashville area.[21] Subsequently, Dall and several others formed the Tennessee Gas and Transmission Company, Incorporated, on April 1, 1940.[22]

Soon after its initial organization, Tennessee Gas's directors sought to increase the size of their company by acquiring other operating entities. Tennessee Gas investigated the possibility of buying a local gas company to become part of a larger proposed long-distance pipeline. On May 4, 1940, Tennessee Gas purchased the Eastern Tennessee Oil and Gas Company from Victor S. Johnson in exchange for 30,000 shares of its first 45,000 stock issue. Johnson was a wealthy businessman, president of the Mantle Lamp Company of America, and holder of the gas franchise for Knoxville, Tennessee. With that franchise under its control, Tennessee Gas had at least one prospective market for its gas supply. The company also added Johnson to its board as the largest single stockholder; his primary role was to be the major financial backer of Tennessee Gas.[23]

Working out of his Manhattan office and the Willard Hotel in Washington, Curtis Dall actively promoted the pipeline plan. One of his first actions was to employ the New York engineering firm of Brokaw, Dixon & McKee to supervise construction. In addition, Dall began gas sales negotiations with E. I. Dupont de Nemours and Company and Phillips Petroleum Company, which also preliminarily agreed to assist in financing the pipeline.[24] Negotiations continued with several other oil and gas companies for both gas supply and assistance with financing.

Dall's seemingly slow progress in arranging for financing and gas supply worried Tennessee's financial backers; the estimated price tag of the line quickly increased from $12 to $20 million.[25] Corporate Secretary John Buckingham wrote to Dall in one of many letters reflecting similar sentiments that "stockholders here are manifesting some anxiety about your activities there. I have assured them that from reports, you are still pleased with prospects."[26] The young company and its officials had much to learn about organizing a regulated pipeline company during war.

Besides arranging financing and contracting for gas supply and sales, the new company required regulatory approval in order to begin construction of the proposed line. Apparently unaware that their pipeline clearly qualified as an interstate pipeline under federal jurisdiction, Tennessee Gas first approached the Tennessee Railroad and Public Utilities Commission for a certificate of public convenience and necessity.

In its application, Tennessee Gas proposed to construct a 20-inch pipeline to extend from Acadia Parish, Louisiana, where gathering lines would connect with four proposed southeastern Louisiana suppliers and ten gas fields, to a point near Brace, Tennessee. From there, two 12-inch extensions were planned to other areas — one to Lebanon, Tennessee, and the other to Knoxville, Tennessee — each to serve numerous communities along the way. The new company, however, feared that United Gas Corporation, the dominant gas distributor in the Louisiana area, might feel threatened by its plans. "We do not wish to get in a wrangle with United Gas on the southern portion of our line," reported one company official; Tennessee Gas did not target southern markets.[27] Substantial markets existed to the north, and the company estimated that it would have an annual capacity of 40 bcf and a potential industrial market of 49 bcf. Tennessee Gas stated that in its first year of operation, it would expect to have the ability to sell 22 bcf at $0.1688 per mcf, and 32 bcf by year five. Any remaining demand would be satisfied by local coal gas production.

During initial discussions with the state commission, Tennessee's directors learned that since they were planning to build an interstate pipeline, they needed a certificate from the Federal Power Commission. The FPC required an applicant to prove during public hearings through extensive expert testimony that it had a twenty-year supply of natural gas for its proposed customers, that it had the engineering plans well worked out in detail, and that it had the financing to build the pipeline.

Despite the rapidly increasing demand in Appalachia for new sources of natural gas, Tennessee's initial application did not meet with favor from the FPC. Lacking solid gas supply contracts and financing, Tennessee Gas officials continually postponed scheduled hearing dates while trying to reach agreements with suppliers and financial institutions. Optimism reigned at Tennessee Gas despite lingering doubts among the stockholders. "With all the demand for fuel, especially gas," Buckingham wrote to Dall, "I don't see how we can fail."[28] But this demand could not be met without gas supply and financing.

On July 22, 1941, after extensive hearings, the FPC dismissed the company's application. Although Tennessee Gas's application suffered from several inadequacies, the company was in trouble for an altogether unexpected reason. The FPC claimed that it lacked jurisdiction to grant a certificate to Tennessee Gas. The agency reported that Tennessee had filed its application under section 7(c) of the Natural Gas Act, which covered the commission's regulatory power over the interstate transmission of natural gas, but the FPC claimed that it lacked jurisdiction to approve the application. The commission ruled that its "jurisdiction to issue certificates . . . exists only when [a company proposes] to construct such facilities for the transportation of natural gas to a market in which natural gas is already being served by another natural gas company." Tennessee could not show that its potential market area constituted one already being served by another natural gas company precisely because the fuel had not previously been available there.[29]

Section 7(c), a curiously worded part of the act, was a legislative obstacle to expansion of the natural gas industry. The provision actually protected the coal and manufactured-gas industries by prohibiting the introduction of natural gas into new markets. Section 7(c) explicitly empowered the FPC to regulate only the expansion of natural gas transmission facilities into areas already receiving natural gas. In 1941, Congress slightly modified the provision by allowing temporary war emergency expansions, but railroad and coal interests opposed any attempts by gas companies to use the emergency provision to expand service permanently.[30]

In a precedent case involving Kansas Pipe Line & Gas Company's certificate application in 1939, the FPC acknowledged its authority to certify expansion programs if the applicant met the necessary requirements. However, the FPC's specific interpretation of the provision remained unclear until it claimed a lack of jurisdiction in the Tennessee Gas case. Importantly, the FPC noted in the Kansas Pipe Line case that "Congress did not intend this Commission generally to weigh the broad social and economic effects of the use of various fuels."[31] Here, the FPC made clear that it would not necessarily consider the social and economic dislocations which might result if and when natural gas displaced manufactured gas or other fuels and their industries. However, section 7(c) would have to be modified to allow the FPC to certify pipelines to serve areas not already receiving natural gas before such interfuel competition would result in the Northeast.[32]

After the FPC dismissed Tennessee Gas's application on July 22, the company plotted a new but more difficult strategy for certification. Instead of applying for a blanket federal certificate, the company planned to acquire a certificate in each of the several states in which it intended to operate. Tennessee Gas refiled an application with the Tennessee commission, which instituted its own hearings and accepted all the testimony and exhibits presented during the FPC hearings.

During the hearings held before the Tennessee commission, representatives of several major industries testified on their need for natural gas at a reasonable price. These industries included the phosphate, pulp, paper, pencil, hosiery, structural steel, aviation, and aluminum industries. The steel and chemical industries professed a strong need for natural gas, and representatives of local utility companies expressed their desire for residential natural gas as well. Not only would it provide a fuel for home heating and cooking as an alternative to coal and coal gas, it would also assist the smoke-abatement efforts going on in various cities. The Tennessee commission observed that gas offered a new and useful fuel to assist and spur industrial expansion especially at a time when the availability of electric power had diminished greatly. The southeastern states had faced an electricity shortage since June 1941 due to the substantial increase in the production of defense products and aluminum in particular. The diversion of electrical power to the aluminum industry proved to be a major benefit of Tennessee's application to sell gas in the area that included the nation's largest aluminum plants at Alcoa, Tennessee, and another at Badin, North Carolina, with other plants proposed for the Muscle Shoals, Alabama, region.[33]

Aluminum production increased dramatically after the outbreak of war in Europe in 1939. In 1939, United States aluminum production was 325 million pounds. The Office of Production Management forecast that U.S. wartime aluminum requirements would reach 1.4 billion pounds by 1942.

The agency estimated that aluminum producers required 14 billion kilowatt hours of electricity to produce such quantities of aluminum. The OPM also noted that by 1942, peak load kilowatts had tripled during the previous four years in supply area no. 20, covered in Tennessee's application, and had already exceeded the dependable capacity of the entire area's requirements; additional fuels for the area were essential.[34]

The Tennessee commission proposed that if certified, the pipeline company would engage in a kind of interfuel cooperation in which natural gas could replace electricity in industrial processes requiring heating or hardening only while electricity could be reserved for applications requiring electrolysis, such as aluminum production. Under this plan, natural gas could be used to produce steels, bronzes, alloys, and other metals and to complement hydroelectric power. The state commission then approved Tennessee Gas's proposal to provide natural gas service in twenty-six counties in Tennessee. However, in deference to section 7(c) and the manufactured-gas, coal, and railroad industries, the Tennessee commission stated that Tennessee Gas would not be allowed to compete with any manufactured-gas distribution service operating with a franchise in any of the twenty-six counties "unless the Commission shall first determine that the facilities of the existing plant, lines, or systems are inadequate to meet the reasonable needs of the public, or that the public utility operating the same has refused or neglected or is unable to make the necessary extensions and additions in order to adequately render the proposed service."[35]

The Tennessee commission then explicitly decried the opposition of coal and railroad interests to the pipeline application. Noting that coal and railroad companies had objected to the Tennessee Valley Authority and other electricity and natural gas projects nationwide, the commission expressed the belief that these interests had blindly attempted to impede and delay progress in the economically important expansion of the natural gas industry. The commission stated that in the early 1940s, industrial progress had as much to do with winning the war as any other business goal.[36]

Despite an overall favorable response to Tennessee Gas's marketing strategy, the commission remained unconvinced that Tennessee had sufficient financial backing for the proposed pipeline, now estimated to cost $23 million. The company supported its certificate application with a financing contract with the investment firm of O'Brien, Mitchell, and Company, but the Tennessee commission ruled on September 11, 1941, that the pipeline had 120 days to present a new financing plan, which, if approved, would lead to certification by the commission.[37]

Tennessee Gas ultimately failed to meet this deadline. The newly created War Production Board hindered Tennessee Gas's application by imposing federal war priorities on all steel projects. Thus, before Tennessee could acquire steel, it had to convince the agency that its pipeline system was vital for national defense. Although the Appalachian region's increasing need for natural gas strained regional production, the WPB did not feel that the emerging gas supply problem merited the immense amount of steel necessary for construction of the pipeline. Tennessee's original promoters had little hope for success. Even if Tennessee could surmount its financing and steel acquisition problems to the satisfaction of the Tennessee commission, it would need to repeat the process in each state in which it intended to operate.

Curtis Dall then employed a new tactic to obtain a certificate. Dall began negotiating with the federal government to have the Tennessee Gas pipeline designated a national defense project. Dall reported to the Tennessee board of directors that if he was successful, the WPB would provide the steel priorities necessary for the construction of the line. In addition, he believed that the federal government, specifically the Reconstruction Finance Corporation, would be obligated to finance the construction of the entire line.[38] Dall's efforts were apparently unsuccessful. He reported back to Tennessee Gas headquarters in Nashville that "our company has run into a good deal of semi-hidden opposition here in Washington. . . . It is my purpose to stay here and blast hell out of the opposition if it is the last thing I do."[39]

Perhaps owing in part to Dall's intensive lobbying, Tennessee Gas's fortunes appeared to improve markedly in early February 1942, when Congress amended section 7(c) of the Natural Gas Act. The original section gave the FPC jurisdiction only in cases where a gas pipeline would enter a market already served by natural gas. The amendment deleted the reference to "market" and required a certificate for all completed interstate pipelines regardless of their market. This version of section 7(c) was based upon the certificate section of the Motor Carrier Act of 1935.[40]

The original version had elicited widespread discontent from both the gas and coal industries as well as natural gas promoters: It "proved unsatisfactory to all concerned." Ironically, coal and railroad representatives led the fight for the amendatory legislation because they were particularly concerned about "their status as intervenors in certificate proceedings."[41] They were especially concerned about the application not of Tennessee Gas but of the Reserve Gas Pipe Line Company, in which the Hope Natural Gas Company had an interest, to construct a 26-inch, $80 million pipeline from Texas to New York City. The Reserve Gas group discovered that under the old 7(c) provision, the FPC limited coal and railroad intervention in certificate applications by gas pipelines proposing to serve markets already served by natural gas. The FPC determined that in these markets, the coal and railroad industry did not have a legitimate claim that a new pipeline would threaten their business precisely because natural gas would have been available there previously. The new amendment, however, gave coal and railroad companies the right to intervene in all cases in which a gas pipeline might expand its facilities, except in the case of limited expansions permitted within the line's existing market area.[42] Following the amendment, coal and railroad interests comprised the major intervenors in gas pipeline certificate hearings.[43] The amendment allowed the FPC to grant certificates to gas pipelines to serve new markets not already being served by another gas pipeline, and it gave Tennessee Gas a second chance to acquire a federal certificate. On February 9, only two days after the amendment, Tennessee Gas filed a second application with the FPC. But the company would have to wait for many months before the FPC would consider its application.

Immediately after filing the new certificate application, Dall both met with, and wrote, J. A. Krug, chief of the power branch of the WPB, to push for the requisite steel priorities. Krug responded to Dall: "As I told you . . . approval or disapproval of this project will depend upon the facts as to the increase in war production made possible from the increased availability of natural gas compared with the diversion of steel and other scarce materials required for construction of the pipeline."[44] During the interim, Dall took a leave of absence from Tennessee Gas without surrendering his financial interest and joined the Air Force, thereby removing himself from a direct role in the company's immediate plans although remaining in regular contract with company management.

Other companies proposing to expand their natural gas service or build a new interstate pipeline began bombarding the FPC with applications (see table 3.2). Prior to the amendment only 16 applications for certificates were pending under section 7(c), and the FPC approved only 4 of them—the others were dismissed by the FPC or withdrawn by the companies. After the amendment, natural gas companies submitted 140 applications under a grandfather clause of section 7(c); 100 were approved.

Table 3.2. FPC Certificate Proceedings after Amendment to the
Natural Gas Act in 1942

Period	Applications Filed	Certificates Issued	Other Status
2/7/42–6/30/42	11	1	10
7/1/42–6/30/43	49	5	54
7/1/43–6/30/44	63	33	52
7/1/44–6/30/45	60	53	85
7/1/45–6/30/46	90	56	110
7/1/46–10/15/46	46	48	99
Totals	319	196	

NOTE: Other Status includes applications still pending at end of period and other dispositions.
SOURCE: FPC, Staff Report on the Natural Gas Investigation, *Administration of the Certification Provisions of Section 7 of the Natural Gas Act* (Washington, DC: GPO, January 1947), 8.

The rush for applications under the grandfather clause related to the provision in the original section 7(c) allowing unauthorized expansion of systems within their market area. The 140 applicants included most of the interstate natural gas companies in existence at that time.[45] In addition, companies filed 319 applications for certificates of public convenience and necessity from February 7, 1942, to October 15, 1946. Of the 319 applications, the FPC granted certificates for 196 companies by October 15 and 8,244 miles of pipeline and 449,695 horsepower (hp).

FPC chairman Leland Olds later described the natural gas situation after the amendment of 7(c) in hearings before the Senate Subcommittee on Appropriations. He stated:

> As a result of the lifting of those limitations practically every natural gas company in the country of any proportion is coming into this commission and stating that the house heating load is growing so rapidly, the space heating load is growing so rapidly, and unless we handle certificate applications promptly, it is going to mean that people throughout these regions served with natural gas today are going to be unable to meet the requirements of heating in the coming winter season.[46]

Indeed, the FPC was so inundated with applications for certificates that it allowed minor additions and expansions without requiring companies to follow the certificate process so long as the provision for the determination of service areas was followed.[47]

COMPETITION FOR APPALACHIAN MARKETS

The amendment to section 7(c) coincided with a recognition by both the FPC and WPB of the growing fuel crisis in Appalachia. Early in 1942, the FPC staff conducted a survey of the Appalachian energy situation at the request of the War Production Board, which had authority to ensure that war production facilities had the necessary fuel to operate.[48] The resulting report described the increasing demand for natural gas in the Appalachian area and the ongoing curtailments of natural gas deliveries due to the lack of supply. The FPC report indicated that curtailments would continue. In response, the WPB requested that the FPC compel a pipeline company to connect the existing Panhandle Eastern Pipe Line system to the Ohio Fuel Gas Company, which served part of the Appalachian region suffering from a shortage of natural gas. The WPB committed to provide the necessary allocation of steel for the construction of a connecting pipeline as approved by the FPC. Ultimately, the FPC had to choose either Panhandle Eastern or Ohio Fuel to build and operate the connecting pipeline system. The required facilities consisted of a 16-inch natural gas pipe extending 88,700 feet from some point on Michigan Gas's (a Panhandle Eastern subsidiary) 22-inch line from Detroit to Ohio Fuel's 16-inch line extending toward Toledo west of Maumee, Ohio.

After the WPB issued its requisite preference rating for materials to build the line, the FPC conducted hearings for the project. The pipe was already being rolled before the hearings began; the only question was which company would build and operate the line. The FPC heard proposals from each company. Panhandle Eastern estimated the job's cost at $394,000; Ohio Fuel estimated $410,815. With proposed looping on the system—the addition of parallel sections to increase capacity—the pipe could deliver 51 mmcf/d on peak days to Ohio Fuel. But representatives from Ohio Fuel, expressing their own misgivings about the project, told the FPC that the company could actually supply all its existing customers without the connecting line. The same officials admitted, however, that extra gas could be used, particularly in the Toledo, Ohio, area.[49]

If a certificate for such a line were to be issued, Ohio Fuel proposed that the FPC grant only a temporary certificate lasting for the duration of the war emergency. Conversely, Panhandle Eastern considered the new pipeline to be necessary not only during the war emergency but for augmenting the "rapidly depleting supply of natural gas in the Appalachian area." Owing primarily to Panhandle Eastern's lower estimated price, overall cooperativeness, and intent to use the pipeline after the war to serve the Appalachian region, the FPC granted a conditional certificate to Panhandle Eastern. Although the FPC recognized that the Appalachian shortage might persist beyond the war, it granted only a five-year certificate. It also required that the pipeline be constructed so gas could flow either way and not interfere with either Panhandle Eastern's existing customers or state and local authority over other natural gas operations. Intervenors from the coal and railroad industries did not oppose the pipeline connection at the hearings. But the National Coal Association pleaded that the certificate not be extended beyond the war emergency so that it could not be used to transport natural gas into the postwar Northeast. The FPC approved the certificate on October 2, 1942.[50]

The Panhandle Eastern case indicated that both the FPC and WPB recognized the severity of the Appalachian shortage. The FPC's willingness to promote a pipeline expansion project into Appalachia encouraged others interested in the northeastern market. In particular, Tennessee Gas's promoters seemed to have been following the progress of Panhandle Eastern's certificate hearings. Soon after Panhandle received its certificate, Tennessee Gas, on November 25, 1942, filed a third amended application for a certificate. The third certificate application included a significant change in the routing of the pipeline in order to satisfy government war planners.[51]

Tennessee Gas recognized now that success hinged on its ability to abide by the WPB's war fuel plans, and Dall continued to lobby for a national defense project designation. He also apparently continued negotiations with Jesse Jones's RFC for financial assistance. At this juncture, Brown and Root, the Houston-based engineering and construction firm headed by Herman and George Brown, and its partner, W. E. Callahan Construction Company, expressed interest in increasing their financial stake in the line. "The Brown-Root Callahan Group," reported Tennessee Gas president Harry Tower, "seems to be very anxious to put up $1,000,000 of working capital when as [*sic*] and if the Federal Power Commission issues the Certificate . . . and we have evidence of major financing from the Federal Government."[52]

Buoyed by increasing private and governmental interest, Tennessee Gas amended its certificate application and proposed to deliver gas only to war industries. The company now proposed to construct a pipeline emanating from near Opelousas, Louisiana, to a northern point near

Ashland, Kentucky, or any other point designated by the government, such as a regional aluminum or chemical plant. Although the company stated that its principal service would be to natural gas companies serving the Appalachian area, it also noted its intention to sell gas to companies along the route of the pipeline. In addition, after the war the company planned to construct a "peacetime extension" line from Brace, Tennessee, to Asheville, North Carolina, thereby creating an expanded version of its originally proposed system.

The FPC conducted hearings on the Tennessee Gas application during the summer of 1943. Appalachian area distribution company executives presented data supporting the claim that their region was truly running out of gas reserves. The president of Hope Natural Gas Company, the large natural gas production subsidiary of Standard Oil Company (N.J.), and the vice president and general manager of the Columbia Gas & Electric Corporation, testified that by the conclusion of the war, neither company would be capable of meeting normal gas sales requirements. Thus, according to representatives of the two largest Appalachian gas-distribution companies, gas shortages would prevent them from meeting demand. The witnesses agreed that the shortages had begun before the war but were now being accentuated and accelerated by it.

Together, the Hope Natural Gas Company, the Columbia system, and their affiliates furnished approximately 60 percent of the total natural gas requirements of the region. Since both companies estimated that they would be short of gas supply, the entire region would suffer. The president of the Hope Company stated that in 1943 his company expected to sell 89.8 bcf, about 3 bcf below expected demand. This lower volume would require curtailment of gas sales to existing customers by 114 mmcf on a projected peak sales day in March 1944. Likewise, the vice president and general manager of the Columbia system expected to have a shortage of 14 bcf in 1944 and a need for 40 bcf of non-Appalachian area gas during the first year after the conclusion of the war. Although these statistics supported the need for a new pipeline system such as Tennessee Gas's to bring in southwestern gas, the immediate shortage had to be dealt with first.[53]

Hope Natural Gas's testimony on its pressing need for new gas supplies lent important support for Tennessee Gas's application. Hope's management, however, soon decided that the most efficient way to bring gas to its system was to build its own line. Hope applied for a certificate to build its own pipeline from the Southwest to the Northeast to serve its distribution system. As a subsidiary of Standard of New Jersey, Hope could rely on financial support from its parent company in such a pipeline venture. Immediately, Hope became a formidable competitor, rather than a supporting witness and potential customer, to Tennessee Gas.

The War Production Board was becoming increasingly anxious to raise the level of natural gas deliveries into Appalachia. During Tennessee Gas's hearings, the WPB issued directive no. 10, which provided for Panhandle Eastern to deliver through its newly constructed connection with the East Ohio Company an additional 50 mmcf/d for the year to the East Ohio Gas Company and a total of 1 bcf to Hope Natural Gas. In addition, the FPC's response to the testimony at the hearings on the fuel crisis included an important statement confirming Appalachia's need for gas. The commission stated that the Appalachian region "embraces one of the most highly industrialized areas in the United States . . . [and] the use of natural gas, both for domestic and industrial purposes, has been substantial for many years."[54] The FPC's own staff engineer presented a slightly less dire scenario for the Appalachian gas situation. He testified that during 1942, the Appalachian region's total gas requirements were 490 bcf and that deliveries were short of that amount by only 1 bcf. Moreover, peak-day requirements of 2.5 bcf/d were short by only 100 mmcf/d. The commission itself estimated that during 1943 there would be a 15 bcf shortage in the Appalachian region and peak-day deficiency of 300 mmcf, figures which did reflect a potentially serious shortage.

The FPC concluded this round of hearings on Tennessee Gas's application on July 5, 1943, and affirmed the need for additional supplies of gas in Appalachia. "It is crystal clear," the FPC stated, "that additional natural gas is needed in the Appalachian region. It follows, therefore, that a realistic view of this situation definitely shows that the public convenience and necessity will be served by the construction and operation of the applicant's pipeline into the area if the additional showing hereinafter referred to is made."[55] Tennessee's application nonetheless remained in serious trouble. While the FPC actively supported a new pipeline to connect southwestern reserves with Appalachian customers, Tennessee remained unable to arrange for either financing or gas supply for its line. The FPC allowed Tennessee Gas sixty days to remedy these deficiencies.

The WPB also increased pressure on Tennessee Gas to attain financing and gas supply. On August 28, the WPB informed the FPC that it had designated the project to bring southwestern gas into Appalachia "as an essential part of the war program," and it declared that it would issue the necessary steel priorities for the applicant receiving a certificate from the

FPC so that steel would be available by the fourth quarter of the current year. The War Production Board required that a pipeline be constructed and in operation for the winter of 1944–45. The WPB stipulated that the certified pipeline would have to place all its orders for pipe no later than October 1, 1943. The WPB wanted a pipeline as soon as possible, regardless of whether Tennessee Gas or Hope Natural Gas built it.

The FPC quickly had to choose either Tennessee Gas or its new competitor, Hope Natural Gas, to build the line. The primary difference between the two proposals was that the Hope line would receive its supply from the proven Hugoton field in north Texas and Kansas, generally known as the Panhandle field. As an established natural gas company in the Appalachian region with impressive financial backing of the Standard Oil Company of New Jersey, Hope seemed to have a greater chance of gaining the FPC certificate. The FPC announced that it would begin hearings on the Hope pipeline plan on September 21.

At this point, Curtis Dall approached Nelson Rockefeller, then employed by the U.S. government as coordinator for inter-American affairs, about Hope Natural Gas's competition with Tennessee Gas. Dall believed that Rockefeller, whose family had once controlled Standard Oil, could help arrange a cooperative pipeline venture with Hope. But Rockefeller insisted that he had left the oil business for government service and had no influence on the oil giant. Dall's persistence finally persuaded Rockefeller to suggest that Dall arrange a meeting with the president of Hope at which some compromise might be reached. Whatever influence Rockefeller might have had, though, did not translate into an agreement between Tennessee Gas and Hope, and the competition continued.[56]

Facing stiff opposition from Hope Natural Gas, Tennessee sought to convince the FPC to approve its application before Hope's hearings began. In their search for a long-term supply of gas, Tennessee's promoters discovered that a Chicago-based investment trust, the Chicago Corporation, owned very large quantities of natural gas on the Texas Gulf Coast near Corpus Christi. Victor Johnson, Tennessee Gas director, traveled to Chicago and negotiated a preliminary gas sales contract with the Chicago Corporation and in turn agreed that Tennessee's line would be extended southward to Corpus Christi. The Chicago Corporation controlled three gas-recycling plants that extracted liquids from its natural gas production in local fields and then reinjected the gas back into the wells. Although the extraction operations were profitable, the Chicago Corporation had long wanted to market its gas supply as well.

The Chicago Corporation was particularly interested in the Tennessee Gas line because it had already investigated the possibility of constructing its own pipeline to the Northeast. In the late 1930s, when the Chicago Corporation began acquiring gas reserves in the Corpus Christi area, another local operator, Clyde Alexander, considered promoting a pipeline from the Texas Gulf Coast into the New York area. He contracted Ray C. Fish, then an engineer with Stearns-Roger Manufacturing Company, to design such a pipeline. At the same time, Alexander convinced the Chicago Corporation to assume a financial stake in the proposed Reserve Gas Pipeline Company. Contracting for gas purchases was not a problem, but arranging gas sales contracts in the Northeast was then difficult. The northeastern utilities "didn't want to have their playhouse disturbed," recalled Alexander, "and they had those artificial gas plants and they were making money and they had been making money for a hundred years and they didn't want anybody coming around there and changing their set-up and they wouldn't even talk about it for a long time."[57] Reserve Gas's frontal assault on the northeastern manufactured-gas industry proved unsuccessful.

After the United States entered World War II, the War Production Board's strict steel limitation requirements on steel availability dashed the hopes of Reserve Gas's promoters and the line was not built. Curtis Dall met with Clyde Alexander in early 1943 to discuss mutual problems and possible solutions in their respective pipeline plans, but they arranged no deals at that time. However, this initial contact likely led Tennessee Gas to approach the Chicago Corporation about a gas purchase agreement later that year.[58]

After successfully negotiating a tentative gas supply contract with the Chicago Corporation, Tennessee Gas hurriedly filed its fourth application for a certificate with the FPC. This certificate, filed on August 23, indicated that the Chicago Corporation would supply gas to Tennessee Gas. But other problems remained. Initially estimated to cost $10 million, the pipeline now carried a $48 million price tag. Part of the increased cost reflected the longer route of the line. In order to obtain the Chicago Corporation's gas on the Texas Gulf Coast, Tennessee had planned to build a 1,156 mile, 24-inch line, instead of a 20-inch line, from a point near the Stratton–Agua Dulce gas field area of Nueces County, Texas, near Corpus Christi.

After Tennessee Gas filed its amended application, the War Production Board sought to ensure the viability of its newly acquired reserves. E. Holley Poe, director of the PAW's Natural Gas and Natural Gasoline

Division, estimated that the gas fields contained approximately 3 tcf of gas. Poe noted the relatively minimal current demand for that gas and added that about one-third of the field's gas supply was currently available. "We therefore recommend," wrote Poe, "that reserves and availability of gas in the Stratton–Agua Dulce field are adequate to meet the requirements of the pipe line proposed to be constructed by Tennessee Gas and Transmission Company."[59] This recommendation was an important step in resolving one of Tennessee Gas's two remaining problems.

Tennessee Gas's increasingly improving outlook began to transform it into an attractive investment. According to one well-placed source, Brown & Root, a Texas-based construction company, attempted to expand further its equity in Tennessee Gas. Tennessee Gas promoter Clyde Alexander heard that in early September, Brown & Root representatives met with Jesse Jones in Washington seeking a financial interest in the proposed line.[60] Brown & Root's founders, Herman and George Brown, were among the most powerful of Houston's business elite and comprised part of the so-called 8-F Group, prominent local businessmen who regularly met in Herman Brown's suite 8-F of RFC Chairman Jesse Jones's Lamar Hotel.[61] The Browns' efforts, however, did not lead them to an equity position in Tennessee Gas.

Tennessee Gas's new certificate hearings were scheduled to begin on September 8, 1943. On September 7, the company contracted to purchase its gas supply for a twenty-five-year period from the Chicago Corporation; of the 3 tcf of gas reserves in the area, only 30 bcf were being produced. In addition to the deal with Tennessee Gas, the Chicago Corporation agreed with Manufacturers Light and Heat Company for deliveries of 50–100 mmcf/d of its gas for the use of the Columbia Gas & Electric System to be delivered through Tennessee.

When the hearings began, Tennessee presented its new gas supply plan along with an amended pipeline proposal that placed the origin of the line at a point near Corpus Christi. The line would travel through Louisiana, Mississippi, Tennessee, and Kentucky to Kenova, West Virginia, where an 18-inch extension would connect to Hope Natural Gas Company at Cornwell, West Virginia. Tennessee's new pipeline system would have an initial total compressor power of 58,000 hp from five compressor stations with ten 1,000 hp units each and one station with eight 1,000 hp units. The company expected its initial delivery capacity to be 207 mmcf/d, of which it would deliver 40 mmcf/d to the Kenova connection and 167 mmcf/d to Hope Natural Gas at Cornwell.

But not all participants at the hearings were in favor of additional pipelines to carry new supplies of natural gas to the region. Representatives of the coal industry, labor unions, and railroads stated that present and future needs of the region could be met by coal. In tandem, these witnesses expressed the idea that "the use of natural gas for industrial and space-heating purposes constitutes a dissipation of the natural-gas resources, and threatens the coal industry with ruinous competition." Although the commission recognized the ability of natural gas to compete favorably with coal, it stated that it did not have the authority to "regulate rates for natural gas sold directly to industrial consumers, which class of gas sales furnishes the keenest competition to the coal industry. Nor does our power to suspend rates extend to indirect sales of natural gas for industrial purposes." In essence, the FPC stated that it had no authority to regulate the interfuel competition that would result from the introduction of natural gas into an industrial region previously supplied principally by coal and its by-products. Overall, the FPC remained unsympathetic with the coal industry. The FPC pointed out to the coal representatives at the hearings that on September 18, 1943, the solid fuels administrator for war announced that "coal production has been unable to keep pace with the expansion of war requirements."[62] Although the coal industry blustered at every expansion of the natural gas industry, it was not capable of alleviating the energy shortage. The FPC dismissed the objections of the coal industry regarding increased natural gas supply for Appalachia.

Tennessee's hearings were going well, but the company's financial health remained uncertain. Its prospects of attaining financing dimmed considerably when V. S. Johnson, Tennessee Gas director and financial backer, unexpectedly died at the Willard Hotel in Washington during the course of the hearings. The company had looked to Johnson to arrange and provide financing for the entire venture, and his death at that critical juncture placed the project in peril.

TENNESSEE GAS CHANGES HANDS

As of mid-September, Tennessee Gas lacked adequate financial backing to pay for the pipeline construction. All previous negotiations with the government and private industry proved unproductive. Tennessee Gas officials now discussed financing plans with the Chicago Corporation, which had a strong interest in seeing its natural gas supplies on the Texas Gulf Coast sold through the Tennessee Gas system. For the Chicago

Corporation, the timing was right to make a strategic decision to enter the interstate pipeline business. It offered to refinance Tennessee Gas under fairly generous terms, but it required complete control of the company in return. The Chicago Corporation offered to pay Tennessee Gas $500,000 to discharge all its liabilities, acquire 90 percent of Tennessee's stock, and finance the construction of the pipeline. It also demanded the resignations of all current directors and officers.

On the morning of September 20, the day before the FPC scheduled the beginning of hearings for Hope Natural Gas Company's alternative application, Tennessee Gas's board met to consider the offer. The board, with some reluctance, agreed to the Chicago Corporation's terms; while their pipeline proposal would now find more favor at the FPC, they would no longer be part of the project. Curtis Dall voted in favor of the offer, but he read a statement at the meeting which reflected his belief that his association with Franklin Roosevelt had ultimately caused the downfall of his original conception of Tennessee Gas. Dall stated:

> Efforts have been made by various people in Washington, by innuendo and covert remark, to damage me personally, and our project, implying that I was trying to use the influence of the President in some way or implying to others that it was dangerous to "play ball" with me or my group, on account of my being his son-in-law. . . . For the further benefit of all those of the new group that will presently run this company, may I state that if I had any influence with the President, which I have *not*, I would, under *no circumstances*, try to use it![63]

After Dall and the original directors resigned their positions, the Chicago Corporation nominated a new set of officers and directors.[64] Principal among them was Gardiner Symonds, a forty-year-old vice president in charge of oil and gas operations for the Chicago Corporation. Symonds had a strong educational and business background. With an A.B. from Stanford and M.B.A. from Harvard, he was one of the new breed of academically educated businessmen chosen to head a new company in a growth industry. His business background in banking did not seem on the surface to be adequate preparation for organizing a gas pipeline company. But his aggressive management style and competitive spirit later characterized him as one of the gas industry's foremost entrepreneurs. As the first president of the reorganized pipeline company, he was responsible for assembling a staff to plan the construction of the line. An extremely aggressive manager, Symonds had a keen interest in dominating the northeastern gas market. He began to plot strategy for the growth of what would later become one of the nation's largest corporations.

Only hours after naming new directors, the Chicago Corporation went before the FPC and reported the recent transaction. The FPC immediately granted an oral certificate to Tennessee to build its pipeline on September 20, 1943, the day before the FPC had scheduled hearings on the Hope Company's competing plan; Hope's application was now defunct. Tennessee Gas's oral certificate became official on September 24.[65]

After gaining control of Tennessee Gas, the Chicago Corporation elicited the aid of Paul Kayser in both financing the pipeline and acquiring additional gas supply. Kayser, a Houston attorney and oil and gas man, was president of both the El Paso Natural Gas Company and Gulf States Oil Company. Gulf States controlled gas reserves in the San Salvador field near the Chicago Corporation's gas reserves. Furthermore, Kayser was an attorney for fellow Houstonian and RFC chairman, Jesse Jones. Kayser's association with Tennessee Gas raised eyebrows after Jones agreed that the RFC would finance a large part of Tennessee Gas's construction costs. The deal was officially arranged when Richard Wagner and Gardiner Symonds of the Chicago Corporation met with Jones, who orally agreed to lend $44 million to Tennessee Gas if the company needed money.[66] Tennessee Gas planned first to finance the line through insurance company investment but wanted an RFC commitment in case those financing arrangements fell through.[67]

The RFC's commitment of $44 million to Tennessee Gas, however, was not carte blanche. Typical of RFC loan policy as developed by long-time chairman Jesse Jones, the agency imposed a great many conditions on the company in return for financing, and these conditions were mailed to Richard Wagner of the Chicago Corporation in mid-November 1943. The loan carried a ten-year term at a 4 percent annual interest rate. The RFC required the Chicago Corporation to inject not less than $2.5 million of its own capital before RFC payments would begin and $1.25 million per year up to a total of $12.5 million during the following ten years. The RFC reserved prior approval of all selections of engineers for the line, construction disbursements, and gas contracts. Of course, the RFC loan would not even be made without prior FPC certification and WPB issuance of the necessary steel priorities.[68]

As Tennessee Gas negotiated with insurance companies to finance the line, these firms did not know about the contingent RFC loan commitment. The Chicago Corporation estimated the total cost of the pipeline to be $47.5 million. After unsuccessfully negotiating alternative financing plans for the additional funds, Tennessee Gas did call for its $44 million RFC loan. RFC records disclose that the agency agreed to make the $44 million loan on February 12, 1944, based on Tennessee Gas's "informal Application No. 1." The background of this particular reference remains unclear.[69] It is not certain exactly what role Paul Kayser might have had in the acquisition of the loan, but his influence was popularly "credited with once saving that imperiled undertaking [Tennessee Gas] by persuading his old friend Jesse Jones to grant it an RFC loan in the New Deal days."[70] Tennessee Gas was then in a position to contract for gas purchases from Kayser's Gulf States Oil Corporation for 10 percent of its natural gas supply requirements.[71]

Now that Tennessee Gas was on the verge of becoming a reality, other existing natural gas companies felt the specter of future competition with the new company. In particular, N. C. McGowen, president of the United Gas Corporation, feared that Tennessee Gas would begin interfering in his company's market area. United Gas, once a part of the huge Electric Bond & Share Company, dominated the Louisiana area gas market and the larger "Gulf South," as McGowen called it. In early October, McGowen and PAW executive E. Holley Poe met with Richard Wagner of the Chicago Corporation "to see him and do something about keeping Chicago and Tennessee out of his back yard." Apparently persuaded that Tennessee Gas was not interested in competing with United Gas, McGowen later disclosed to Tennessee Gas officials efforts by others to prevent the construction of Tennessee Gas.[72]

The entire effort to finance and construct the pipeline was an impressive display of wartime business-government interaction and cooperation. With federal approval to build the line, financing in place, and general industry approval, construction began. On October 1, 1943, the company established its first payroll. A small number of employees later swelled to between 9,000 and 11,000 workers during peak construction periods. Tennessee Gas hired several construction firms to work on the pipeline. The primary contractor, Bechtel-Dempsey-Price, shared the work with Williams Brothers Corporation and Brown and Root. The company broke ground for the pipeline on December 4, 1943, at the Cumberland River in Tennessee. Workers welded the first mainline pipe on January 10, 1944, but severe winter weather conditions prevented rapid progress. By May 1, 1944, only 76 miles of the pipeline had been constructed as rain, rough terrain, and material shortages slowed progress.

Tennessee Gas Transmission Company. Solid black line indicates pipeline route.

As the weather improved with the coming of summer, the pipeline moved forward. After laying 1,200 miles of pipe, securing rights-of-way from thousands of landowners, crossing sixty-seven rivers and hundreds of roads, and building seven compressor stations, Tennessee Gas began delivering fuel through the pipeline on October 31, 1944 (see map).[73]

Tennessee's primary gas market area was the Appalachian region. By the end of 1945, its first full year of operation, Tennessee had delivered 73.5 bcf of gas into Appalachia. Tennessee began operations with five Appalachian area distribution companies, which had a combined customer base of 750 Appalachian area communities. Approximately 95 percent of Tennessee's early gas sales, though, went to the two large Appalachian distribution systems: United Fuel Gas Company, a subsidiary of Columbia Gas & Electric Corporation, and its former competitor, the Hope Natural Gas Company, now a subsidiary of Consolidated Natural Gas Company. Tennessee thus became the first pipeline to connect Gulf Coast natural gas reserves directly with Appalachia.[74]

After the Tennessee Gas pipeline was constructed and in operation, the RFC released its interest in the pipeline. It sold its $44 million interest in the line to the Chase National Bank of New York for $45,157,863.[75] To pay off the bank note held by Chase National, Tennessee offered a $57.5 million package to the SEC, which quickly approved it. The plan included the sale of $35 million in First Mortgage pipeline bonds, stocks, and a $15 million loan. Chase National, First National of Chicago, Continental Bank & Trust, and Harris Trust & Savings of Chicago all participated in the plan, which allowed Tennessee Gas to pay off the original loan. At about the same time, Howard J. Kessner resigned as a director of the RFC and took the position of vice president and director of the Chicago Corporation. Although the RFC no longer owned an interest in the line, two of its former associates were now directors of the company.[76]

Tennessee Gas's success, however, caused other problems for its parent company as well. During December 1944, the FPC instituted an investigation to determine if the Chicago Corporation through its ownership of 81 percent of Tennessee Gas stock qualified as an interstate natural gas company as defined by the Natural Gas Act of 1938. If so, the FPC could investigate whether any rates, charges, or classifications relating to any of the Chicago Corporation's natural gas operations were also subject to FPC jurisdiction. Not wanting to risk an unfavorable FPC ruling, the Chicago Corporation began divesting itself of Tennessee Gas stock. First, the parent company considered the possibility of acquiring an additional loan for an extension of its line or paying the existing loan off early. Jesse Jones reported to the RFC board that Richard Wagner had phoned him to discuss these options. It is unlikely that Wagner truly desired another RFC loan for Tennessee Gas since the Chicago Corporation had already begun divesting itself of Tennessee Gas stock. Jones reported that he told Wagner Tennessee Gas could "of course, pay it at any time according to the prepayment privileges which, for the present, would be 104 and accrued interest. I told him," Jones said, "I felt sure the Corporation [RFC] would not be willing to sell the loan on any different basis."[77]

The Chicago Corporation divested itself of Tennessee Gas after a syndicate of underwriters purchased the stock. Stone and Webster, the lead underwriter, became Tennessee Gas's controlling stockholder. By September 4, 1945, the Chicago Corporation had no financial stake in Tennessee Gas. At the same time, the Chicago Corporation's representatives on the board of directors, including Paul Kayser, resigned. Gardiner Symonds, the firm's forty-two-year-old president, severed his ties with the Chicago Corporation and retained his executive post. Symonds was not only the chief manager of the company, he quickly stood out as the dominant force on its board of directors as well. An attorney from Tennessee Gas's outside law firm later remarked that Symonds "was a

Table 3.3. Tennessee Gas Transmission Company Operations, 1945–1954

Year	Gas Sold and Transported (bcf)	Operating Revenues ($ million)	Miles of Pipeline	Personnel	Reserves (tcf)
1945	74	14	1,284	634	na
1946	95	18	1,309	1,099	na
1947	109	20	1,630	2,020	4
1948	155	28	2,561	1,661	6
1949	221	41	3,048	1,993	8
1950	286	53	4,294	2,680	10
1951	386	76	5,859	2,752	10
1952	454	106	6,755	2,979	12
1953	495	133	7,277	2,916	12
1954	506	143	8,177	3,300	12

SOURCES: Tennessee Gas Transmission Company, *Annual Report* (various years), and Moody's Public Utility Manual.

taskmaster, there's no doubt about that, but he was a very fair man."[78] Tennessee Gas was now for the first time an independent corporation with a single strong leader and certain future as a major interstate pipeline company (see table 3.3).[79]

Tennessee Gas continued to function effectively and expanded its system capacity to meet the increasing Appalachian demand. The War Production Board requested Tennessee Gas to "give consideration to the feasibility of installing additional compressor stations, compressors, and auxiliary apparatus" on its natural gas pipeline system to alleviate ongoing shortages.[80] After the WPB's request, Tennessee filed an application with the FPC for a certificate of public convenience and necessity to expand its system. The expansion program called for the lease of four additional compressor stations to be constructed and owned by the Defense Plant Corporation, additions to its current compressor stations, and construction of ninety-five miles of 16-inch outside-diameter pipe from the San Salvador gas field in Hidalgo County, Texas, northward to Tennessee Gas's pipeline in Nueces County, Texas. The addition of the compressor stations would increase Tennessee Gas's delivery capacity by 60 mmcf/d.[81]

During the FPC's hearings on Tennessee's application, several expert witnesses again questioned the actual severity of the Appalachian shortage. It became evident that representatives from two of Tennessee's largest customers, Consolidated Natural Gas and the Columbia system, had greatly overestimated their respective expected deficiencies of natural gas. The Columbia system representative stated that his company would have a shortage of 169 mmcf/d out of total peak-day deliveries of 1 bcf/d in the winter. But the FPC, after considering all Consolidated's testimony and sources of supply, stated that the shortfall actually could be no more than 30–45 mmcf/d. In addition, the FPC concluded that the Consolidated system would have a 30 mmcf/d shortage out of total peak deliveries of 651 mmcf/d only if its peak days occurred in February; otherwise, no shortage was expected. Despite the evidence that Consolidated had intentionally overestimated its shortage of gas supply, the War Production Board sent a letter to the FPC urging an increased flow of natural gas into the Appalachian region to fuel production for the duration of the war.[82]

The letter swayed the commission to approve a limited version of Tennessee Gas's request. The commission did not approve the construction of the San Salvador project, ruling that this extension's projected capacity of 94 mmcf/d would be used only at 29 percent capacity.[83] Moreover, the extension would not collect additional gas for Tennessee, it would only replace a portion of Tennessee's supply currently purchased from the Chicago Corporation. And this supply was in no immediate danger of diminishing. The FPC denied that Tennessee Gas could lease the four compressor stations from the Defense Plant Corporation (DPC), which it planned to purchase after the war to increase its capacity by 60 mmcf/d. The commission also determined that Tennessee should not be able to charge its intended rate of $0.2175 per mcf. Based on a 6.5 percent return and the costs of the proposed additions to existing compressor stations and the calculations used to determine rates, the commission calculated that Tennessee could charge no more than $0.1825 per mcf. The commission then issued a certificate for the modified project covering only the period of the war emergency.[84]

FPC commissioner Leland Olds issued a strongly worded dissenting opinion. A trained utility economist, former head of research for the AFL's Railroad Employees Department, and outspoken critic of much of the FPC's own policy, Olds rejected his fellow commissioners' eagerness to permit such large-scale expansion in the gas industry. He argued that a serious shortage of natural gas did not exist in Appalachia. Olds stated that "cross examination of company witnesses also determined that the commission was being asked to give its assent to what was in fact only a limited segment of a broader plan for postwar deliveries of Texas gas to northeastern markets in much larger quantities." Foreseeing the potential for a dramatic introduction of southwestern natural gas into the northeastern states, Olds commented that as the war emergency production requirements tapered off, the commission was being "crowded" with applications for natural gas facilities which "taken singly, do not appear too significant but cumulatively might seriously dislocate the balanced utilization of the country's energy resources."[85] Olds warned his fellow commissioners that Tennessee's application should be viewed in the context of the unprecedented growth of the gas industry, a process Olds hoped to slow to a more manageable pace. His interest in strictly regulating the industry made him no friend of natural gas executives.

In particular, Olds distrusted Gardiner Symonds's motives. During the hearings, Symonds indicated that Tennessee Gas was positioning itself to increase its sales capacity for the postwar period. Olds was concerned about the potential for energy shortage during the war, and he did not believe that any wartime energy adjustments should be used as a springboard from which a natural gas company could conquer the northeastern coal and manufactured-gas industry after the war. Olds believed that Tennessee's application was masked to hide its true intent of preparing to increase deliveries of Texas-produced gas into "the coal producing Appalachian area and the industrial Northeast." And he feared that if the commission inadvertently established the basis for the modern natural gas industry in the northeastern states, the effects of the conversion from coal to natural gas might not be adequately prepared for or regulated. Olds concluded his dissent: "Thus I do not find in the record any sufficient evidence of war need to warrant action on the present application without full investigation of the impact of all plans for sale of southwestern gas in the Northeast. We cannot deal with the situation piecemeal and at the same time conserve the public interest for which the Congress made us responsible in 1942 when it amended section 7 of the Natural Gas Act."[86]

Olds's objections to the continued rapid expansion of the natural gas industry fell on the unsympathetic ears of both his fellow commissioners and the gas industry. Natural gas operators clearly understood that expansions to their systems allowed during the war would put them in an advantageous competitive position after the war to expand faster than other newly formed competing companies. The WPB and FPC generally supported expansions based on current demand without significant concern for postwar interfuel competition. Thus, Tennessee Gas was able to maneuver its wartime expansion strategy to position itself for future expansion. Panhandle Eastern had been in a similar situation, but it was less well situated to expand into new northeastern markets because it already served large and growing midwestern markets.

Tennessee Gas began as an entrepreneurial effort focused on selling southwestern gas into gas-short Appalachia. With the outbreak of war, the pipeline venture required intensive governmental assistance for its success. The WPB provided the steel necessary for its construction, the RFC financed it, and the FPC certified its operation. However, Tennessee Gas was a private business, albeit one operating in a heavily regulated wartime industry. Its promoters accepted strict government controls during the war in anticipation of a postwar boom in the natural gas industry. Shepherded through the war by the FPC and the special powers of other war emergency agencies, Tennessee Gas prepared to become the only pipeline capable of serving the vast northeastern markets.

The war years witnessed comprehensive federal involvement in the gas industry, ranging from financing pipeline construction to the strict allocation of gas sales. The FPC and the WPB worked together, although not always harmoniously, to alleviate the Appalachian fuel shortage. They provided more stable sources of natural gas for war industries, thus meeting their primary goal. With the assistance of the various federal agencies involved in its creation, Tennessee Gas survived several changes of corporate control. Importantly, its leaders recognized from the beginning that if their company could secure a place in the wartime natural gas industry, it would be in an excellent competitive position to break into the large northeastern markets after the war.

World War II offered the natural gas industry an opportunity to expand in return for allegiance to strict regulatory control. In particular, intense Appalachian demand for natural gas stimulated entrepreneurial interest in the gas pipeline business. Also, the amended section 7(c) of the Natural Gas Act allowed natural gas pipelines to apply for certificates to serve areas not already served by natural gas. This was an immensely significant change in the 1938 act. The regulatory modification encouraged intense competition among natural gas companies for FPC certification to serve new markets. And it opened up the capabilities of pipelines to transport natural gas to the Appalachian area war industry. It also provided the future opportunity of selling natural gas to markets farther to the east after the war, and it gave coal, railroad, and other interests a greater opportunity to present opposing views to the FPC regarding gas industry expansion.

The Appalachian gas shortage compelled the FPC and other agencies to look favorably upon plans by pipeline companies to bring natural gas into the Northeast. The FPC joined the WPB and RFC in imposing a high level of central planning on the industry's expansion. Panhandle Eastern's expansion and Tennessee Gas's construction clearly resulted from wartime demand compounded by declining regional supply. The FPC determined when, how, and at what price these systems would operate. At this time, the FPC acted to some extent as a governmental pipeline promotion agency, and this biased activity appeared to be in response to the war emergency. With FPC certification, pipeline companies could legitimately expect RFC funding and WPB steel priorities. Government assistance, if not intervention, was a necessary component in the wartime expansion of the natural gas industry.

After the end of the war, federal control of the industry gave way to market-driven expansion. Several other proposed and expanding pipeline companies also targeted the huge northeastern metropolitan areas of Philadelphia, New York City, and Boston, among others, which remained without natural gas. An intense competition among these new lines for gas pipelines and sales into the Northeast was about to begin. Appalachia, as it would turn out, was a timely stepping stone, reached with the assistance of a wartime regulatory regime, to the Northeast.

4. War Pipelines and Peacetime Markets

WORLD WAR II boosted energy demand while hastening the construction of pipeline systems capable of transporting fuel to vital consumption centers. Wartime energy shortages spawned new oil and gas pipeline construction between the Southwest and Northeast. But American wartime energy requirements placed a far greater burden on the oil industry than on the natural gas industry. As the lifeblood of the war, oil attracted intense government and industry attention. The United States was the primary supplier of both fuel and war material to the Allies, who required tremendous amounts of petroleum to build and power an assortment of equipment from bombers to tanks. To ensure the flow of fuel for the war effort, the United States government supported increased petroleum production and financed the construction of five petroleum pipeline systems.[1]

Two of the government's war emergency pipelines, the Big Inch and the Little Big Inch, were particularly vital to the war effort. These were the longest petroleum pipelines ever constructed. The 24-inch-diameter Big Inch extended 1,254 miles from the East Texas oil fields at Longview, Texas, to Phoenixville, Pennsylvania, with extensions to points near New York City and Philadelphia. The 20-inch-diameter Little Big Inch originated in the refinery triangle area of Port Arthur and Houston and extended to Linden, New Jersey (see map). The Big Inch carried crude while the Little Big Inch transported refined products.

During the war, the Inch Lines became heroic symbols of the American energy industry when they replaced a crippled oil tanker fleet. Early in 1942, German submarines began sinking U.S. oil tankers transporting petroleum from the Gulf Coast up the Atlantic seaboard. The submarine

The Big Inch and Little Big Inch pipelines. Solid black lines indicate pipeline routes.

attacks, and threat of attack, virtually halted the ocean-going oil transportation system and threatened the East Coast and the Allies with an oil shortage. "We tried to keep that [shortage] a secret," recalled one government official. "It was a necessary effort not to let the Germans know how much damage they were doing but we just couldn't move the oil."[2] During 1942, oil tanker shipments from the Gulf Coast to the New York harbor area fell from an average of 1.5 million barrels per day to approximately 75,000 per day.[3] Quickly, and with RFC financing and WPB steel priorities, eleven major oil companies under contract with the government built the Inch Lines to alleviate the northeastern oil shortage. Together, the Inch Lines delivered more than 350 million barrels of crude and refined products during the war. In evaluating the contribution of the Inch Lines to the war effort, one petroleum pipeline historian wrote: "It would be difficult to measure what the availability of this amount of petroleum and petroleum products meant to the war effort, both overseas and at home. But clearly it was a major contribution."[4]

THE INCH LINES AND PUBLIC POLICY

The postwar fate of the Inch Lines, the only pipelines that directly connected the Texas oil and gas region with the Philadelphia and New

York City area, quickly became the focus of a "classic clash of interest groups."[5] Large oil companies argued that the Inch Lines were unnecessary for peacetime petroleum transportation systems and should be converted to natural gas. Some smaller independent petroleum firms believed that the Inch Lines could be successfully operated as oil lines, but the industry as a whole rejected this idea.[6] The coal and railroad industries uniformly objected to the possibility of converting the Inch Lines to natural gas carriers. They believed that the introduction of natural gas into the Northeast would displace the sale of both coal and manufactured gas and generally disrupt the coal industry. Railroad companies which transported coal for the coal companies likewise feared that a drop in coal demand due to the introduction of natural gas into the Northeast would disrupt the railroads as well.

Even before the war ended, however, several major oil companies actively supported the concept of the postwar conversion of the Inch Lines to natural gas to keep them out of petroleum transportation. Sidney A. Swensrud, vice president of the Standard Oil Company (Ohio), was one of the most vocal proponents of this view. Early in 1944, he presented what was essentially an oil industry policy statement on the Inch Lines. In a paper entitled "A Study of the Possibility of Converting the Large Diameter War Emergency Pipe Lines to Natural Gas Service after the War," Swensrud echoed the concerns of his industry: "The disposition and future use of these lines constitutes, in the minds of many people, one of the most important problems of post-war readjustment in the Oil Industry."[7] Swensrud concluded that the Inch Lines could successfully and economically transport natural gas from the area of the nation's greatest natural gas reserves in the Southwest to the densely populated Northeast without causing a serious disruption in the coal industry. At the heart of the matter, however, was the fact that peacetime transportation of petroleum by oil tanker was expected to be less expensive than by pipeline. Further, no oil company wanted to see a competitor controlling the two Inch Lines, which might potentially monopolize the oil transportation industry. Thus, Swensrud was so intent on keeping the Inch Lines out of petroleum service that he and J. Howard Marshall, former chief counsel and assistant deputy administrator of the PAW and president of the Ashland Refining Company, secretly bankrolled a public relations effort favoring the conversion of the Inch Lines to natural gas.[8] The debate over the Inch Lines continued and intensified after the conclusion of the war.

During late 1945, the RFC discontinued wartime operations of the Inch Lines, and the governmental process of determining their postwar future began.[9] The Inch Lines, along with all other war surplus property, became subject to the jurisdiction of the Surplus Property Act of 1944, which required the Surplus Property Administration (SPA) to determine the best disposal policy for all war surplus property. The act stipulated that the SPA could not sell any property when a monopoly might be created, encouraged the SPA to develop new businesses through the sale of surplus property, specifically mandated the sale of all surplus government-owned transportation facilities in order to promote an adequate and economical national transportation system, and required the SPA to obtain for the government the fair value of the surplus property at the time of the sale. In addition, the act gave the United States government the legal right to reacquire and operate the property in the event of another national emergency.

As the demobilization process gained momentum, war surplus property disposal and policy functions were separated into different agencies. On November 9, 1945, the War Assets Corporation began administering the disposition of certain categories of war surplus property including all property, such as the Inch Lines, officially owned by the Department of Commerce and the RFC, while the SPA retained jurisdiction over disposal policy. In order to make its own recommendation as to the disposition of the Inch Lines, the RFC hired the engineering firm of Ford, Bacon & Davis, Incorporated, to study the options for their postwar use. The firm's report, issued on August 15, 1945, advised that both Inch Lines should be converted to natural gas transmission. Although this consulting report was not binding, it did provide an initial assessment of the postwar commercial value of the Inch Lines. Soon after receiving the report, the RFC began its own public inquiry on September 14 by mailing telegrams to 135 users and potential users of the pipelines and requesting general inquiries or informal bids for the purchase or lease of the lines.[10]

At the same time, Senator Joseph C. O'Mahoney (D-WY) prepared a series of special hearings to consider how to dispose of the government's vast array of war surplus petroleum property. The hearings lasted for three days, from November 15 to 17, 1945, and they provided a forum for representatives of competing industries and interests to speak publicly about the disposition of the Inch Lines.[11] A total of twenty-eight representatives of government agencies, railroads, oil, gas, and coal interests presented oral and written statements. Opinions regarding the Inch Lines generally followed self-interest and represented vastly different views: natural gas representatives supported a government sale of the lines to a gas company, coal and railroad groups suggested scrapping the lines or using them for oil transportation only, and major oil companies stated the lines were not needed for a peacetime oil pipeline network, but independents wanted them in petroleum service so that they could have access to the large markets without using pipelines controlled by the major oil companies.

The first witness at the hearings was W. Alton Jones, president of the War Emergency Pipelines and chairman of the Committee on Postwar Disposal of Pipe Lines, Refineries, and Tankers. In discussing options for peacetime use of the Inch Lines, Jones acknowledged that they had raised controversy even before they had been built. In response to questions, Jones admitted that some oil and railroad companies contributed to the initial delays in planning for the construction of the Inch Lines by insisting that they have no postwar economic use. Jones also recalled that during some of the early planning conferences for the construction of the Inch Lines, some participants discussed the possible peacetime use of the Inch Lines for natural gas transportation. In discussing the development of a policy to guide their sale, Jones made four recommendations: (1) Because of their military value the lines should not be scrapped; (2) the lines should not be utilized in the postwar period for the transportation of crude oil or refined products either to the East Coast or to the interior; (3) under certain conditions, it would be feasible to convert the lines to natural gas service; and (4) in the event that the lines were not converted to natural gas service, they should be removed from use by congressional direction and held as a military asset.[12]

Several other persons testified on their interest in acquiring the Inch Lines. Charles H. Smith, a veteran of the coal industry, told the congressional committee that he represented a group of persons interested in purchasing the Inch Lines for natural gas transmission. He described himself as a mining engineer and consulting public-utility engineer with a company in New York. As the chief consulting engineer for the Ordinance Department and for the Bituminous Coal Administration during the war, Smith recognized the opportunities offered by the Inch Lines for natural gas service. Smith was not insensitive to the needs of the coal industry, and he stated that the loss of manufactured-gas markets to natural gas and the resulting decrease in coal demand might be mitigated by an increased exportation of American coal to Europe. Smith described the extensive destruction of European coal mines during the war, and he believed that postwar Europe would depend on the American coal industry for supply. Smith also argued that the Inch Lines would provide work in the gas fields for returning soldiers and other unemployed people. He believed that natural gas, a less expensive boiler fuel than fuel oil or manufactured gas, would increase production of finished products in New York. According to Smith, these finished products would increase railroad shipments, yield more employment opportunities, and offset any resulting railroad coal transportation reductions.[13]

Another witness supporting conversion to natural gas was Claude A. Williams, a forty-one-year-old Austin-based attorney representing his uncle, Rogers Lacy, who owned large natural gas reserves located in East Texas. Williams had been the assistant secretary of state in Texas between 1938 and 1940 and the chairman and executive director of the Texas Unemployment Compensation Commission from 1940 to 1945, when he began working with his uncle to market natural gas. Williams and Lacy were part of a group of gas reserve owners, which also included the Alfred C. Glassell family, planning to form a corporation to buy the Inch Lines and convert them to the transportation of natural gas produced from Texas fields. Williams's plan was to operate the Inch Lines as common carriers so that any producer could transport natural gas to the Northeast on a pro rata basis depending upon the carrier's natural gas production and the capacity of the Inch Lines.

Williams recognized the increasing pressure on politicians to address the issue of Texas flare gas. Many oil drillers valued gas only as a medium to provide the necessary pressure in oil wells to lift the oil through the drill casing to the surface, and these drillers typically either allowed the gas to escape into the atmosphere or flared it at the wellhead. Referring to the tremendous volume of waste gas, Williams began his testimony with a humorous story. He told of an Irish admirer of Patrick Henry who was unable to speak at a meeting. Instead, he gave a note to the chairman that said, "I regret very much that I am unable to speak to you folks tonight because I seem to have a lot of gas on my stomach, and, as Patrick Henry would say, it is crying out, 'Give me liberty or give me death.' "[14] With this story as a preface, Williams argued that there was simply too much natural gas in the Southwest to continue to neglect or waste as oil companies had been doing. Williams stated that as much as 1 bcf/d of natural gas was being flared instead of shipped to market. Citing the success of the Tennessee Gas Transmission Company in selling gas produced on the Gulf Coast in West Virginia, he said that southwestern gas could be used by eastern consumers. If the Inch Lines were not sold for natural gas use, he maintained, gas companies would no doubt build other natural gas pipelines following the same route.

E. Holley Poe, the former AGA and PAW gas official, also spoke in favor of converting the lines to natural gas. Poe had left the PAW sixteen months earlier to form a natural gas consulting firm called E. Holley Poe and Associates. After leaving PAW but before starting his own firm, Poe was the executive vice president of the Petroleum Reserves Corporation, a government agency under the RFC designed to investigate foreign oil supplies. Poe did not stay with the PRC long and may not have impressed his colleagues there. In one meeting at which he was absent, the minutes of the PRC reported that "although it was generally agreed that the present Executive Vice President of the corporation had done a highly creditable job, . . . the corporation should obtain . . . a man of outstanding ability who would not be objectionable to the industry and who would command public confidence."[15] In any case, Poe's entrepreneurial leanings and expert knowledge of the gas industry had led him to the conclusion that the Inch Lines offered both opportunity and profit.

Poe did not mention at the hearings that he was also actively engaged in a plan to acquire the Inch Lines. He, internationally famous geophysicist Everette DeGolyer, and Norris McGowen, president of United Gas Corporation, the largest gas distribution company in the Southwest, had been following the developments surrounding the Inch Lines for at least two years. McGowen's United Gas controlled the largest southwestern gas distribution system and was a major gas producer—capable of supplying the Inch Lines with gas for northeastern markets.[16] Poe presented a forceful statement on the great potential for natural gas transmission through the Inch Lines. He showed that a potentially huge market for natural gas existed in the northeastern states. Using studies done by Everette DeGolyer, Poe argued persuasively that this market could be served by the tremendous natural gas reserves in the southwestern states. Poe claimed that the Inch Lines could be operated at about 80 percent capacity and supply the major metropolitan areas in Pennsylvania, New Jersey, and New York. Cognizant of the coal industry's fear of natural gas, Poe proposed that natural gas be used only to enrich manufactured gas so that no coal would be displaced. The enriching process normally used water and petroleum, and Poe estimated that as many as ten million barrels of enriching petroleum per year would be displaced by natural gas.

But spokesmen for the petroleum industry did not object to losing this business. Even if some residential coal gas was displaced, Poe argued, "out of a total of some fifty-five million or fifty-six million tons of bituminous coal and seventy million tons of anthracite coal that goes into the area, you will probably find less than three percent goes into the manufacture of gas." Poe did not address the effects of direct competition between natural gas and coal for industrial markets. Instead, he suggested that natural gas and coal companies could coexist in northeastern markets. But Poe also spoke bluntly about the coal industry's lobbying efforts to block competition from natural gas. "I cannot be too concerned," he said, "about the position of the railroads and of the coal people. I do not think that any producer of a commodity has any inherent authority over any market so that he can make his market buy his product at a higher price than it should be sold for."[17]

Coal and railroad interests expressed adamant opposition to converting the Inch Lines to natural gas. Their basic argument concerned the potential for a loss of sales and jobs. Both F. C. Wright, Jr., representative of anthracite producers, and W. K. Hopkins, general counsel for the United Mine Workers of America, voiced fears that natural gas transported through the Inch Lines would displace 6 million tons of coal and translate into a loss of 11,650 jobs based on a calculation that each miner averaged 505 tons of coal production per year. These 11,650 unemployed workers would then draw about $780 per year in government relief for a total of $9.1 million annually. Hopkins was more general but equally emphatic. The Inch Lines, he said, "should not be leased or sold to private industry or be governmentally operated or subsidized. They should be held intact in reserve against any future national emergency. . . . [Oil or natural gas transportation through Inch Lines] would certainly tend to seriously interrupt and impede the post-war reconversion and recovery."[18]

O. E. Schultz, chairman of the Coal and Coke Committee, Trunk Line Territory, quantified the potential damage to the railroad industry resulting from a loss of coal load due to a conversion of the Inch Lines to natural gas. He said that natural gas would displace at least 20 percent, or seven million tons, of the annual coal traffic handled by the railroads to the Philadelphia and New York City markets. This coal displacement would translate into a loss of railroad industry gross revenue of up to $16,450,000 annually, which would convert into a loss of 3,306 jobs. James M. Souby, general solicitor of the Association of American Railroads, presented a similar argument.[19]

After the conclusion of these hearings, the Federal Power Commission began its own intensive hearings on the state of the natural gas industry. Known as the Natural Gas Investigation, docket G-580, these hearings took place throughout the nation between October 9, 1945, and August 2, 1947, with the last hearing in Washington, D.C. The FPC designed these hearings to elicit industry opinion on practically all aspects of the natural gas industry to assist it in administering the Natural Gas Act and to propose new legislation to improve the overall quality of the industry if that proved necessary. These hearings provided a new forum for those supporting, and opposing, use of the Inch Lines for natural gas transportation, but they only indirectly affected the fate of the Inch Lines.[20]

Having reviewed various studies and testimony from the O'Mahoney hearings, the Surplus Property Administration prepared in early January 1946 to issue its disposal policy for all surplus property including the Inch Lines. The SPA published this policy on January 4, 1946. The SPA report, commonly referred to as the Symington Report after SPA administrator W. Stuart Symington, disappointed natural gas interests as well as major oil companies by advising that the Inch Lines should be sold for petroleum transportation. The report noted that the pipelines would foster low-cost and dependable petroleum transportation for the small independents, which did not have access to either pipelines or tankers, and the report favored allowing a newly created petroleum company to operate the Inch Lines. It concluded that the Inch Lines "would be vital in the event of another emergency," and a "careful consideration of all the factors involved leads to the conclusion that the Big Inch and Little Big Inch should be kept in petroleum service." The report addressed the issue of converting the Inch Lines to natural gas and agreed that this was a viable possibility, but it found that to do so would greatly lessen the value of the lines, which were designed to transport petroleum.[21] This conclusion pleased the coal and railroad industries, slighted the proposals from natural gas interests, and threatened the major oil companies.[22]

Representatives of major oil companies responded by letter to Senator O'Mahoney with a sharp attack on the Symington Report. Calling the SPA's recommendations "unsound and not in the public interest," the letter stated that only under ideal peacetime conditions (which would be impossible to attain) could the Inch Lines be economically operated as petroleum carriers. In addition, the letter observed that "it is most unfortunate that the SPA report implies that the lines were originally advocated by the industry with a view to post-war transportation of petroleum. The opposite is true."[23] The authors went on to recommend either maintaining the Inch Lines in an idle state for use in a future emergency or allowing them to be used for natural gas transportation. If used for natural gas, the letter continued, petroleum fuel oil, not coal, would be displaced in the production of manufactured gas on the East Coast. The letter, though, had no influence on SPA policy.

Yet the Symington Report did not diminish the interest of several groups in using the Inch Lines to transport natural gas. Two companies led by persons who had testified at the O'Mahoney hearings submitted informal bids for the Inch Lines. Charles H. Smith had organized two companies for this purpose. One of these, Big Inch Oil, Incorporated, offered $40 million for the Little Big Inch to carry petroleum; and the other, the Big Inch Natural Gas Transmission Company, offered $40 million for the Big Inch to transport natural gas. Claude Williams, who along with partners had formed Transcontinental Gas Pipe Line Company in February, also offered $40 million for both Inch Lines to ship natural gas. The War Assets Administration (WAA) did not act on these or several other informal bids whose terms were not made public. The high value and widespread interest of various companies dictated that the SPA would have to provide a public forum for the sale of the Inch Lines, and the SPA had six months' time to begin the sale process.[24]

Soon after the SPA issued its disposal policy report, President Truman issued an executive order consolidating surplus property functions into a new agency (the WAA). Truman then transferred the functions of the SPA to the War Assets Corporation and then merged the War Assets Corporation into the War Assets Administration, effective March 25, 1946. Significantly, the WAA was an agency within the Office for Emergency Management of the Executive Office, responsible to the president. The WAA, under its chief, Lieutenant General E. B. Gregory, now had singular responsibility for further policy questions and disposal of more than 90 percent of existing surplus property.[25]

THE FIRST AUCTION

The WAA prepared to sell the Inch Lines under the Symington Report's guidelines. On June 7, 1946, six months after the publication of the Symington Report, and immediately after the RFC officially declared the Inch Lines surplus property and provided for their transfer to the War Assets Administration, the WAA announced an auction for the Inch Lines. All bids would be due by July 30 and be opened and read the following day. The WAA advertised the Inch Lines for sale or lease nationally in thirty-eight newspapers and five oil trade journals. The advertisements described the government's disposal policy as recommended by the Symington Report: "In accordance with the disposal policy indicated in the Surplus Property Administration Report, first preference will be given to continuing the Big and Little Big Inch in petroleum service, thereby assuring availability of the lines in the event of a national emergency." The disposal policy stated that "special attention will be paid to offers which would give the many small and independent petroleum operators the opportunity of participating in both the use and acquisition of this facility in whole or in part. As the SPA report suggests 'this might be accomplished by the creation of a corporation, the shares of which would be held by various individuals, cooperatives, or associations.' "[26]

Although the advertisements also invited bids for converting the Inch Lines to natural gas, the WAA made clear that gas bids would receive lowest priority. The coal industry as a whole was pleased by this. Representative Walter of Pennsylvania, who had spearheaded an effort by the coal industry and labor unions against converting the Inch Lines to natural gas, also reported to the news media that Brigadier General John J. O'Brien, WAA deputy administrator, promised him that if no petroleum bids were adequate, the WAA would not act on the natural gas bids until Congress could be consulted. To this, Walter said, "I've won my fight. If bids from any of the oil companies are acceptable and the lines are continued in oil transportation, that's fine. Otherwise, we still have a chance to stop their conversion into natural gas."[27] Walter represented Pennsylvania's official position that the Inch Lines should not transport natural gas into the coal-rich state. However, natural gas entrepreneurs continued preparing to bid on the Inch Lines, and large northeastern utilities began studying the possible effects of replacing some of their gas-manufacturing capacity with gas transported via the Inch Lines.[28]

The WAA received sixteen valid bids by the auction closing date of July 30. During a public hearing on July 31, WAA personnel read out loud each of the bids (see table 4.1).[29] Of the bids submitted, seven proposed oil transportation, three were for gas, five proposed some combination of oil and/or gas use, and one did not state a preference. After all the bids were in, the WAA began the process of determining the winning entry. This proved to be more difficult than expected. The WAA had imposed no standards on the format for the bids, and this situation created serious

Table 4.1. Bids for the Big Inch and Little Big Inch, July 30, 1946

Bidder	Proposed Use	Terms	Price ($ million)	Cash at Closing ($ million)
Big Inch Natural Gas Transmission Co.	gas	sale	85	85
Transcontinental Gas Pipe Line Co., Inc.	gas[1]	sale or lease[2]	85	85
E. Holley Poe & Assoc.	gas	sale or lease	80 or[3]	80
Glenn H. McCarthy	gas	sale	80	80
Big Inch Oil, Inc.	oil	sale	110	66
J. W. Crotty & Assoc.	gas[1]	sale	127.5	5
Mutual Oil-Gas Transmission Co.	oil	sale	70	—
W. Lee Clemons	oil[1]	lease[2]	rent[3]	na
L. M. Glasco	oil	lease[2]	rent[3]	na
Frank M. McCraw	oil	lease[2]	rent[3]	na
J. H. Moroney	oil or gas	lease[2]	rent[3]	na
Noyack Oil Corp.	oil	sale	80	—
Russell Palmer	oil or gas	sale	135 (oil) 115 (gas)	—
Ryford Pipeline Co.	oil	sale or lease[4]	30 or[3]	—
Sinclair Refining Company	oil	sale or lease	open	—
Syndicated Industries, Inc.	oil	lease[2]	—	—

NOTES:
[1] or both oil and gas.
[2] lease with purchase option.
[3] rent to be determined by rental formula.
[4] purchase of Little Inch and lease with purchase option of Big Inch; purchase of Big Inch at $40 million.
SOURCE: "Analysis of 'Big' and 'Little Inch' Big Proposals," War Assets Administration worksheet.

problems, particularly in evaluating the financing of each bid; it was difficult to compare the bids in terms of their dollar values. While most of the bidders offered some cash for the pipelines, they proposed that the remaining cost would be covered by various forms of debentures, payments based upon gas sold, or other arrangements. In addition, the highest up-front cash amounts for the Inch Lines came from companies proposing to use the Inch Lines to transport natural gas.

The WAA decided to consider only the cash portion of the bid as the total bid price. After defining its pricing criteria, the WAA listed the highest bids as those proposing to use the Inch Lines for natural gas transmission. Both the Transcontinental Gas Pipe Line Company, led by Claude Williams, and the Big Inch Natural Gas Transmission Company bid $85 million in cash. The group led by E. Holley Poe, who had also testified at the hearings, had the second highest cash bid for natural gas of $80 million, after the WAA reduced its total bid of $100 million to $80 million, the amount of cash offered.

Charles H. Smith, representing Big Inch Oil, Incorporated, submitted a bid to use the Inch Lines to transport oil, which most closely met the WAA's surplus property disposal policy as described in the advertisements. But his cash price of $66 million was substantially lower than those for natural gas; of a total bid price of $110 million, he offered $44 million in indentures. At the November hearings, Smith had stated his intention to purchase the Inch Lines for natural gas. But his bid was for using the pipelines as petroleum common carriers. Smith's proposal stated that the pipelines would transport oil for the many small producers in Texas. Since the lines would remain in petroleum service, they would be ready for oil use in time of national emergency. The most unusual bid came from L. M. Glasco, who proposed to dismantle the Big Inch and then rebuild it from the Permian Basin in West Texas to California. It would then operate as a crude oil pipeline.

A few days after the WAA received the bids for the Inch Lines, it announced its priority in analyzing types of bids based upon the Symington Report's recommendations.[30] The order of preference of consideration was (1) oil, (2) oil and gas combinations, (3) gas, and (4) any other uses. Also, the WAA stated that bidders would be expected to produce substantiating data for their bids as the bids were being analyzed. One week later WAA notified each bidder to submit these extra data by September 9, 1946, a date later extended to September 16. The WAA continued to request additional information from bidders without coming close to reaching a decision.

Indicating the intensely political nature of the bid for the Inch Lines, almost all the bidding companies included high-profile former, and sometimes even current, political figures. Many powerful men had aligned themselves with various groups of investors looking to acquire the lines. Washington insiders lobbied for the various groups, and the lobbying reached a fever pitch. One bidder, Big Inch Gas Transmission

Company, included a former Ohio senator, an ex-justice of the Supreme Court, and an ex-chairman of the Maritime Commission. Attorneys for this company were listed as former Roosevelt aide Thomas Corcoran, a former general counsel for the FPC, and a former FPC commissioner. Another bidder, American Public Utilities, included a distinguished pair of Washington lawyers, former trustbuster Thurman Arnold and one-time undersecretary of interior for Harold Ickes, Abe Fortas. Maverick oil man Glenn H. McCarthy rounded out the list.[31]

Another prominent bidder, E. Holley Poe, had also aligned himself with a cadre of powerful business and political figures. Poe originally worked with geologist Everette DeGolyer in formulating a bid for the Inch Lines. In later discussions with Charles Francis, a Houston attorney with the firm Vinson, Elkins, Weems, and Francis, at the FPC's broad-ranging "Natural Gas Investigation" in which they were all involved, Francis agreed to work with the group to bid on the Inch Lines. These men soon recruited Reginald H. Hargrove, a vice president of United Gas Corporation, with the blessings of his boss, N. C. McGowen, to assist the group; at one point McGowen had taken a preliminary financial interest in the venture but soon decided that if United Gas controlled the Inch Lines, United Gas would be in violation of the PUHCA.[32] The final and most important additions to the group were George and Herman Brown, of the Houston-based construction firm Brown & Root, who offered both financial backing and engineering support. Earlier in the century, Herman Brown had taken over a small road-building company from his former employer and, with the help of his younger brother George, turned it into a large construction company. Herman Brown was the rugged contractor who firmly controlled the company from the inside while George, the taller and more outgoing of the two, parlayed his contacts in both business and government into big construction contracts. The brothers were not new to dealing with federal agencies. They had built hundreds of destroyer escorts at their Brown Shipbuilding Company near Galveston during the war and participated in the construction of the Corpus Christi Naval Air Station.[33]

The Brown brothers were well connected in both business and political circles. In Houston, they were part of the 8-F Group, which regularly met in Herman Brown's suite in the Lamar Hotel, owned by RFC chairman Jesse Jones, who had provided financing through the RFC for both Tennessee Gas and the Inch Lines. In particular, the Browns were ardent supporters and friends of Lyndon Johnson. Francis later reported that he solicited Johnson's aid for his group's bid, but there is no evidence that Johnson assisted the group.[34] Nevertheless, Donald Cook, who subsequently became a highly trusted adviser to Johnson, helped complete research for the Poe group.[35]

Another prominent Washington figure, George E. Allen, did associate with the Poe group. A former hotel operator and commissioner of the District of Columbia, Allen was a trusted friend, adviser, and speech writer for President Truman. Truman appointed Allen to the position of director of the RFC in January 1946. Through Frank Andrews, president of the New Yorker Hotel and cousin of Reginald Hargrove, Allen met several times with members of the Poe group to discuss its plans to bid for the Inch Lines.[36]

While the various groups attempted to promote their bids, both overtly and covertly, Harold Ickes took a journalistic shot at the Poe group. At the time, Ickes published a syndicated column, "Man to Man." He wrote an article about the Poe group entitled " 'Uncle Jesse's' Bid for Oil Pipelines Seen Loaded Two Ways for Monopoly."[37] In it, Ickes suggested that former RFC chairman Jesse Jones was behind the "so-called E. Holley Poe bid" for the pipelines. Ickes warned that "if the story is true, the War Assets Administration had better do some keen sniffing because Uncle Jesse has no peer among horse traders that I have known." Ickes was more certain that George Butler, a Houston lawyer, "husband of Jesse's only heir and custodian of many of Jesse's enterprises, both business and political, is one of the E. Holley Poe crowd." Ickes noted, however, that he liked Poe's bid.

Ickes's article set off a flurry of denials from Poe, Francis, and Butler. Poe wrote to Ickes and met with him two days later, assuring him that Butler and Jones were in no way connected with the bid. Francis did the same, and Butler issued a statement denying that he was part of any group interested in the Inch Lines. Ickes wrote to both Poe and Francis: "I cannot see that the column did your group any harm. On the contrary, it served notice on Jesse Jones that he was more or less suspect and he might be disposed to keep his hands off. That is really what I had in mind in writing the column."[38] Although there is no evidence that Jones ever had a direct interest in Poe's bid, his close business relationship with the Brown brothers evoked suspicions that he may have been involved with the company's early history.

Although Ickes apparently satisfied himself that he had at least warned Jones and Butler to stay clear of the Inch Lines, his article set off alarm bells at the WAA. The agency requested from all bidders, the day after Ickes's article appeared, a list of the "identity and business connections" of all individuals or firms associated with the particular bids.[39] Poe responded to L. Gray Marshall of the utilities branch of the WAA that "my proposal and supplementary data discloses the names and official connection of all persons associated with me."[40]

But Ickes's initial attack on the Poe group led to a correspondence with Poe and Francis that was important to their efforts. Ickes wrote that he believed the Inch Lines should be used to transport natural gas, and he assured Francis that he would soon write a column to that effect. Ickes subsequently reaffirmed in his columns that the Inch Lines should be used for natural gas and not oil, and his opinion helped the Poe group in its efforts to have the WAA change its policy of selling the Inch Lines with top priority for oil transportation. Ickes wrote in his column that using the Inch Lines to transport natural gas would "end John L. Lewis' stranglehold on the economy of the United States."[41] As president of the United Mine Workers, Lewis had ordered the nation's coal miners to strike during the war and consequently aggravated energy shortages in Appalachia and along the eastern seaboard. The specter of future disruptions of energy supply by prolonged and often violent coal miners' strikes was a powerful incentive for government officials to reconsider their policy of favoring oil over natural gas in the sale of the Inch Lines.

Amid the political intrigue and public concern over the bid, the WAA was in a quandary. Despite the Symington Report's recommendation to use the Inch Lines to transport oil, political pressure from supporters of natural gas such as the Texas Railroad Commission, Harold Ickes, and major oil companies had stalled the process of selling the Inch Lines. To assist in evaluating the various proposals for the pipelines, the WAA created a Special Advisory Council, which met to discuss the bids for the Inch Lines. This group included representatives of federal departments and agencies involved with oil- and gas-related matters. The council met on October 9 and 14, but it reached no consensus on how to dispose of the Inch Lines. Two council members, the Army-Navy Petroleum Board and the Department of the Interior, nonetheless issued important statements supporting the conversion of the Inch Lines to natural gas.

The prime consideration of the Army-Navy Petroleum Board was that the Inch Lines would be available for petroleum shipments in the event of a national emergency, but it voiced no objection to a sale of the Inch Lines for natural gas use as long as the government could recapture the pipelines for petroleum use within ninety days of a national emergency. The Department of the Interior offered indirect support for the conversion to natural gas by expressing concern over the tremendous quantities of flare gas wasted in Texas which might otherwise be transported to markets by the Inch Lines. Without a clear consensus, the WAA invited the Federal Power Commission to join in the planning for the disposal of the Big and Little Big Inch Pipe Lines. Aware of the intense political controversy swirling around the issue, the FPC declined this invitation and stayed out of the bidding process.[42]

By late October and early November, several articles in trade journals and daily newspapers cited rumors that the WAA was ready to award the Inch Lines to a bidder for oil transportation. In one of these articles, Elliot Taylor, editorial director of the journal *Gas*, claimed that the WAA was on the verge of awarding the Inch Lines to Charles H. Smith's Big Inch Oil, Incorporated, presumably because this bid corresponded most closely with the SPA's surplus property disposal policy.[43]

The Poe group also heard the rumors. William H. Leslie, an aide to George Brown, wrote Brown of a conversation with J. Ross Gamble, a Washington-based attorney who represented the Poe group. Gamble reported that "it is extremely likely that the Big Inch Oil, Inc. bid will be recommended for approval unless our group takes early and aggressive steps to prevent this happening." Leslie continued to say that if a sale of the Inch Lines would not be made for natural gas transportation, "he thinks that steps should be taken to encourage members of congress to request or instruct that congress be given a chance to investigate the matter more fully to determine what usage the lines should be sold for. This might result in a delay which might give our group time to fight the opposition more successfully by bringing public and congressional opinion around to our way of thinking."[44] Publicly, the Poe group responded to the rumors when Poe told the media that the WAA would be responsible for the continuing waste of natural gas in the Southwest if the pipelines remained in oil service.[45]

As the WAA wallowed in confusion, the supporters of a sale of the Inch Lines to a gas bidder received an unexpected boost. John L. Lewis assisted their cause when he began threatening to call another nationwide coal strike. Lewis's threats sparked an immediate public backlash against him, and this sentiment encouraged support of a sale of the Inch Lines for natural gas transmission. Both Harold Ickes in his syndicated column and Marshall McNeil flailed Lewis for threatening to strike. Ickes suggested that a good dose of natural gas competition via the Inch Lines would serve Lewis right. And Ickes chastised the WAA for simply not selling the Inch Lines to the highest bidder. Anything else, he wrote, "is just plain politics."[46]

Lewis was apparently unconcerned about inadvertently supporting a conversion of the Inch Lines to natural gas. On November 20, he declared void the contract between the government and his coal labor union of some 400,000 members to which they had mutually agreed during the previous May. The resulting strike proved short-lived, as Lewis, after the government fined both him and his union, called off the strike on December 7. Numerous newspaper articles and editorials accused Lewis of creating a national fuel shortage after having insulated himself from its effects. "Lewis' house at Springfield won't be cold on account of the coal strike," the *Washington Post* reported. "The United Mine Workers chief had gas heating installed in June, 1945."[47] At the same time, Harold Ickes accused John R. Steelman, a close Truman associate and the reconversion director, of conspiring with Lewis in trying to delay the disposition of the Inch Lines. Steelman denied the charge.[48] But the publicity surrounding Lewis's actions and the growing fear of increased fuel shortages in Appalachia gave the WAA an added rationale for favoring the conversion of the Inch Lines to natural gas.

Early in November, Big Inch Oil, Incorporated, the bidder rumored to be the likeliest winner of the bid, filed a letter with the WAA stating its willingness to help relieve the shortage of natural gas in Appalachia. Tom Corcoran of the Big Inch Natural Gas Transmission Company told an attorney for the Poe group that he would soon do the same. The combination of the coal strike and the Appalachian shortage gave the gas bidders an increasing advantage in the bidding war.[49]

The failure of the advisory board to reach a conclusion on the sale of the Inch Lines portended a renewed congressional investigation into the Inch Line question, something gas bidders wanted all along. On November 19, the Select Committee of the House of Representatives to Investigate the Disposition of Surplus Property (referred to as the Slaughter Committee after its chairman, Roger C. Slaughter) met with the sole purpose of determining the best way to dispose of the Inch Lines. These hearings became a forum for the continued debate on the Inch Lines. Thirty-nine individuals, representing all facets of the energy industries, testified concerning the disposition of the Inch Lines. Fifteen of the individuals who testified supported a sale for natural gas use; ten favored oil transportation; two wanted one of the pipelines to be used for natural gas and the other for oil. Some speakers argued that the Inch Lines should not be used for natural gas or oil, but they gave no alternative. Once again, the hearings illustrated the intense competition generated by the Inch Lines among different industries.

THE FIRST AUCTION IS VOIDED

On the opening day of the hearings, General Robert M. Littlejohn, chairman of the WAA, cleared the air by announcing a decision to reject all previous bids for the Inch Lines. Littlejohn, who had been WAA administrator since July 22, 1946, and previously held the post of chief quartermaster of the European Theater of Operations, had been frustrated over the months of delay in selling the Inch Lines. His decision to reject the previous bids was an attempt to start the process over again on new terms. Espousing the philosophy of his West Point wrestling coach, "There ain't no hold that can't be broke," Littlejohn was ready to break the Inch Line logjam.[50] In support of Littlejohn's decision to throw out the bids, the interagency committee studying the disposal policy stated: "It had become evident that the interest of national defense could be met regardless of whether the pipe lines be used for natural gas, petroleum and its products, or a combination thereof . . . the bids had been invited on a restricted basis, which precluded the government from securing the maximum net cash return."[51] Since the highest cash bids had been those for converting the Inch Lines to natural gas, this statement suggested a giant step away from the policy called for by the Symington Report. In fact, the Symington Report had been largely discredited after Stuart Symington testified at a congressional hearing that one of his young and inexperienced aides who had virtually no knowledge of the oil and gas industries wrote most of the report. Littlejohn also announced that the WAA had estimated the fair value of the Inch Lines to be $113,700,000 as of September 30, 1946, a figure significantly greater than the highest cash value bids previously received.[52]

John L. Lewis's coal strike, combined with an increasingly severe energy crisis, prompted the various government agencies to seek a role for the Inch Lines to alleviate the energy shortage. Several bidders and Tennessee Gas discussed with the WAA the possibility of temporarily leasing the Inch Lines to alleviate the Appalachian fuel shortage. Other participants in the meeting included representatives of Big Inch Natural

Gas Transmission Company, Big Inch Oil, Incorporated, and Transcontinental Gas Pipe Line Company. Claude Williams, president of Transcontinental, also expressed interest in leasing the Inch Lines during a radio interview on "Headline Edition" on November 26. These interested parties met with representatives of the Department of the Interior, FPC, Army-Navy Petroleum Board, and War Assets Administration to determine how best to utilize the Inch Lines to alleviate the Appalachian fuel shortage. Tennessee Gas was the only operating gas pipeline and consequently gained the edge against the others.[53]

Tennessee Gas had been interested earlier in acquiring the Inch Lines, but the company decided after making a thorough study of them that it had its hands full with its own Appalachian service area; Tennessee Gas did not join the sixteen other firms that did bid on the Inch Lines on July 31. Although the Inch Lines extended through the Appalachian region, they also passed through Pennsylvania and into New York; they were well situated to serve customers in the northeastern market area. But Tennessee Gas, said its president Gardiner Symonds, "had so much to do to meet the needs of our companies in the Appalachian area that we could not, until we had completed the construction of all the facilities we sought — and we have since sought additional facilities — have any interest in operation or ownership of the Big or Little Inch lines to seek a new service area."[54]

LEASING THE INCH LINES

However, Tennessee Gas's Washington attorney, Robert May, discussed with the WAA in September the possibility of leasing the lines. On October 23, Tennessee Gas had formally offered to lease and interconnect them with its own system in order to serve the Appalachian region. The company proposed only to lease the Inch Lines temporarily to alleviate the Appalachian fuel shortage.[55] The WAA responded that as it was still in the midst of analyzing the previously submitted bids for the lines, it could not process Tennessee Gas's proposal.

On November 29, the company made a formal proposal to lease the Inch Lines on a temporary basis and deliver gas to Appalachia — but not farther eastward. In its letter to the Department of the Interior, Tennessee Gas described its ability to act promptly and disclosed its willingness to have the FPC direct the disposition of the natural gas it would transport. A cover letter dated November 30, 1946, described Tennessee Gas's purpose for leasing the Inch Lines: "TG&TC desires to emphasize the fact that the sole purpose of its proposal is to attempt to effect a temporary connection of these lines with the existing facilities of the Tennessee Company in order to meet the emergent need for natural gas in the Appalachian area. This proposal covers only a short term lease and does not contemplate permanent operation of these lines."[56]

The proposal for an interim lease identified four major Appalachian gas companies as the primary recipients of the natural gas: Ohio Fuel Gas Company, Manufacturers Light and Heat Company, Kentucky Natural Gas Company, and East Ohio Gas Company. Up to 120 mmcf/d was currently available for transport, and Tennessee Gas was to receive $0.24 per mcf and pay the government $0.06 per mcf sold. Tennessee Gas also agreed to spend $250,000 for material and installation involved in converting the lines for natural gas use and connecting them to Tennessee's own lines.

On December 2, the WAA accepted Tennessee's proposal and issued the company a letter of intent to operate the pipelines. Presumably, Tennessee received the letter instead of the other interested companies because it was an operating gas company which could most quickly and efficiently put the Inch Lines into service. The lease began at 12:01 A.M. on December 3, 1946, and was to expire at midnight on April 30, 1947.[57] Tennessee Gas had insisted on a lease lasting at least through April to prevent it from losing money on the project. The ever-present Harold Ickes, in both congressional testimony and in his syndicated column, condemned the lease agreement. He characterized Tennessee Gas as attempting to monopolize all natural gas shipments to the Northeast, and he also accused the company of pulling political strings and preventing a possible competitor from even temporarily operating the Inch Lines.

Tennessee did interconnect its system with the Inch Lines near Many, Louisiana, and began pumping gas into the Little Big Inch and the Big Inch on December 5 and December 9, respectively. By December 11, gas was flowing to consumers in Ohio. However, the company did not actually convert the pumping stations along the Inch Lines to transmit natural gas; instead, it simply operated them without compressor stations by using only the 140-pound pressure exerted by the natural gas emerging from the various connecting wells. Tennessee's operation of the Inch Lines proved their viability for transporting natural gas and seemed to give that company an apparent edge in the next round of the bidding. Not to be outdone and uncertain of the fate of the Inch Lines, Claude Williams then applied to the FPC for a certificate of public convenience and necessity to construct a new pipeline to the Northeast just in case he might be unsuccessful in winning the bid for the Inch Lines.

The FPC's mounting concern over the energy shortage in Appalachia also prompted it to impose emergency service rules on December 12, 1947, for the Panhandle Eastern Pipeline Company. Panhandle Eastern's pipeline system extended from the gas fields of the Texas Panhandle and southern Oklahoma into large markets in Missouri, Illinois, Indiana, Ohio, and Michigan. Under its broad powers to regulate the interstate transmission of natural gas, the FPC required Panhandle Eastern to supply gas on an emergency basis to certain customers in Appalachia, and to curtail all its interruptible industrial customers so that residential users requiring space heating would not freeze during the winter.[58] The interim lease to Tennessee and the emergency service rules imposed on Panhandle Eastern clearly indicated that the Inch Lines were needed to deliver natural gas to the Appalachian region.

WAA administrator Littlejohn reconsidered the disposal policy for the Inch Lines, surely stinging all the while from constant high-profile criticism for inaction by Harold Ickes and others. On December 18, he issued three basic recommendations for a revised policy on the sale of the Inch Lines: (1) The Inch Lines should be disposed of for natural gas or oil; (2) all pumping and other oil equipment should be maintained by the purchaser so government could recapture the pipelines within ninety days of a national emergency; and (3) the sale should be made to the purchaser offering the greatest net return in dollars to the United States government.[59] These new provisions included additional challenges to the bidders. If the oil-pumping stations had to be maintained by the purchaser, the natural gas bidders would have to determine for themselves whether the stations could be kept in their pipeline's rate base. Otherwise, the company could not factor the cost of the stations into their gas sales rates, and the stations would then be a wasted expense. The provision calling for the greatest net return tended to undermine the Surplus Property Act's concern for competition and the Symington Report's focus on national security, but it did not preclude an oil company from submitting a winning bid. Again, the consideration of the winning company's rate base was critical. It was unclear whether the depreciated value of the lines on the government's accounting books, the original cost of the lines, or the value of the bid plus conversion costs would form the new rate base.[60]

THE SECOND AUCTION

Now convinced of the viability of using the Inch Lines for natural gas transmission, Littlejohn announced on December 27 a new bid for the Inch Lines. All bids were to be submitted to the War Assets Administration, Office of Real Property Disposal, Industrial Division, P.O. Box 2707, at the Washington, D.C., post office. The bids would be made on standardized forms, including a section requiring each bidder to detail its plans for the utilization of flare gas, to the WAA at noon on Saturday, February 8, 1947. To purchase the Inch Lines, the successful bidder would be required to make four payments over a nine-month period. The first payment of $100,000 would accompany the actual bid as a deposit. A second payment of $1 million would follow the WAA's issuance of a letter of intent to sell the pipelines to the high bidder. The third payment of $4 million would be due on the day the bidder began operating the Inch Lines. The winner of the bid would make the final payment to be collected and transferred to the WAA to complete the purchase within nine months of the date of the letter of intent.

Littlejohn sent his official report to Congress on January 3, 1947, and stated his intention to sell the Inch Lines for either oil or natural gas. Congress then had thirty days to consider the WAA's recommendation that the Inch Lines be sold for either the transportation of oil, its products, or natural gas or a combination of them.[61] On the same day and in response to Littlejohn's report, Representative Francis E. Walter of Pennsylvania introduced a congressional resolution, H.J. Resolution 2, to prohibit the WAA from selling the Inch Lines until six months after the final report of the FPC's major Natural Gas Investigation (docket G-580). Walter, who represented Pennsylvania's anthracite coal industry, adamantly opposed the use of the Inch Lines for natural gas. He feared that the coal industry would not survive competition with natural gas for northeastern fuel markets.

The House Committee on Interstate and Foreign Commerce then held hearings to discuss the Walter resolution. In order to consider H.J. Resolution 2, the House Committee on Interstate and Foreign Commerce directed the FPC to provide as soon as possible a report on the natural gas industry. The FPC produced the report, titled *Statement on Natural Gas for the House Committee on Interstate and Foreign Commerce with Reference to H.J. Res 2 on Disposition of the Big Inch and Little Big Inch Pipelines*, on January 23, 1947. It described the tremendous natural gas reserves in the

Southwest, the outrageous volumes of gas (1 bcf/d in Texas and 2 bcf/d in the Southwest as a whole) flared or vented into the atmosphere each day, as well as the growing demand for natural gas in the Northeast.

In a last-minute attempt to persuade Congress to adopt his resolution against increasingly difficult odds, Walter discussed the economic effect the Inch Lines as natural gas pipelines would have on the railroad industry. He argued that if converted to natural gas, "the Inch Lines would transport an annual equivalent of about five to six million tons of coal . . . [and if] the densely populated Eastern Seaboard is afforded natural gas service for domestic, industrial and commercial purposes, the quantities necessary to meet the demands of that area will result in the displacement of enormous quantities of solid fuel." Walter continued:

> The competitive threat to the railroad industry as a result of the conversion of the Big Inch lines to natural gas is alarming. Anthracite coal is the very life blood of many of the railroads operating in the area in which my District is situated. The Eastern Seaboard is by far the greatest market for anthracite and a displacement of several millions of tons of this type of fuel could not but have an extremely serious effect upon what are generally termed the anthracite railroads. Bituminous coal is the fuel almost exclusively used for industrial purposes on the Atlantic Seaboard and in New England. A displacement of substantial tonnage of this type of fuel would have seriously detrimental repercussions on that segment of the railroad which supplies the transportation which brings bituminous coal to the eastern region.[62]

However, Walter's statements changed few, if any, minds.

Congress did not adopt Walter's resolution for several reasons, including the practical considerations that further delays in the sale of the Inch Lines would inevitably mean that the WAA would continue to incur an expense in maintaining the lines and probably receive a lower price for them when they finally would be sold. The rejection of still another six-month delay also suggested that Congress was growing weary of the protracted political in-fighting over the Inch Lines and was ready for this much publicized issue to be resolved one way or the other.

The debate over the Inch Lines continued, but it was clear that natural gas use was becoming an increasingly attractive option to both the WAA and even Congress. Not only would a natural gas bid bring the government a higher return on its investment in the Inch Lines, it also would open up a part of the country long dependent upon a troublesome fuel, coal, to an abundant supply of natural gas. On February 3, Congress cleared the

WAA to sell the Inch Lines for oil or natural gas use only five days before the February 8 deadline for the second round of bids.

Now, natural gas interests seemed in control of the process. The important element for each of the bidders was simply outbidding the others. Interested bidders engaged consulting engineers and Wall Street bankers to help them determine the value of the Inch Lines. Tennessee Gas, which was currently operating them under a temporary lease, had in its possession all the Inch Line accounting books and directed a team of accountants to "comb them out real thoroughly, . . . [to learn] exactly what the replacement cost of that system would be."[63] With the aid of Stone & Webster, Tenneco's largest stockholder, Tennessee Gas settled on a bid of $123,127,000, which reflected the estimated depreciated cost of the Inch Lines, not their value as natural gas transporters.

Claude Williams was also preparing a second bid for the Inch Lines. He submitted to the WAA a bid of $131 million, only $1 million more than the estimated cost of an alternative gas pipeline that his company, Transcontinental Gas, proposed to build from Texas to New York. But Williams had originally planned to bid much higher. Working with two accountants from Kuhn, Loeb & Company, the Williams group earlier decided on a bid of $148 million only to have Kuhn, Loeb chairman, Elisha Walker, threaten the night before the bids were due to pull his firm's financial backing out of the venture if Williams submitted such a high bid.[64] Williams's group also felt pressure from its competitors. Charles Francis of the Poe group suggested to Transco days before the bid that the two companies submit a joint bid, but Clyde Alexander, a Williams associate, recalled that "all of the time I think that he was trying to find out how much we were going to bid."[65]

The E. Holley Poe group had reorganized specifically to bid on the Inch Lines. They renamed their company Texas Eastern Transmission Corporation, with George Brown as chairman and Poe as president. Texas Eastern contracted the services of Dillon, Read and Company, and August Belmont IV, vice president of the firm, worked with the group. Great-grandson of August Belmont, the American correspondent banker for the European Rothschild family and founder of the Belmont Stakes, Belmont had a name that evoked an aura of high capitalism. A small and energetic man, Belmont earned great respect for both his in-depth knowledge of investment banking and ability to communicate complicated transactions to the press. The Texas Eastern group worried about the regulatory constraints of the bid. Belmont recalled that Texas Eastern's promoters were unsure whether the FPC would allow it to "earn 6 percent on what you paid for the line, or whether you could only earn 6 percent of what the actual physical stuff on the books was."[66]

On the night before the bid, Brown, Poe, Francis, Hargrove, and August Belmont met to determine their bid. Belmont came to the meeting after agreeing with Douglas Dillon, chairman of Dillon, Read, on top figures for Texas Eastern's bid. Belmont convinced Dillon that "the project could stand a price of $140 million" and asked Dillon to give him permission to suggest a $135 million bid, with authority to go as high as $140 million. Before the Texas Eastern night session was over, the group decided to bid $143,127,000, and Belmont received Dillon's approval for the higher amount the next day.[67]

On Saturday, February 8, all bids were due at WAA's post office box. August Belmont, Ted Wadsworth, J. Ross Gamble, and George Pidot of Shearman & Sterling, Texas Eastern's New York attorneys, delivered the bid to the War Assets Administration office located at the U.S. post office.[68] To prove that they had actually deposited their bid at the post office and on time, they had a photographer take their picture on the post office steps. Gamble convinced the postmaster general, Robert E. Hannegan, who was a personal friend, to allow him to sit in the post office gallery and watch box 2707, in which the bids were being deposited, in case anyone attempted to open the box and view or remove any of the bids.[69] By the end of the day, the WAA had received ten bids.

On February 10, WAA representatives opened each of the bids in a public hearing; five were accepted and for various reasons the other five were considered void (see table 4.2). The three highest bids received by the WAA were for natural gas use. Texas Eastern had submitted the high bid. The amount of $143,127,000 was only $2.5 million less than the Inch Lines' original construction cost of $145.8 million, and it was more than $12 million above the second-highest bid of $131 million submitted by Claude A. Williams and Associates representing the Transcontinental Gas Pipe Line Company. Tennessee Gas and Transmission's bid came in third at $123,700,000. The two other bids accepted by the WAA were from Big Inch Oil & Gas Corporation for $108,031,660 and Big Inch Natural Gas Transmission Company for $121,000,000. The low bid was submitted by J. W. Crotty of Dallas. He had pasted a dollar bill on the bid form, $0.60 for the Big Inch and $0.40 for the Little Big Inch. Crotty noted that the enclosed funds were all his, and he would be the principal executive, or in his words, "the whole cheese."[70]

Table 4.2. Bids for the Big Inch and Little Big Inch, February 8, 1947

Name	Proposed Use	Terms	Price Offered
Texas Eastern Transmission Corp.	gas	sale	$143,127,000
Transcontinental Gas Pipe Line Co., Inc.	gas	sale	131,000,000
Tennessee Gas Transmission Co.	gas	sale	123,700,000
Big Inch Natural Gas Transmission Co.	gas	sale	121,000,000
Russell Palmer[1]	oil or gas	sale	120,000,000
Big Inch Oil & Gas Corp.	oil or gas	sale	108,031,660
George J. Callahan[1]	oil	sale	6,100,000
Dr. John Bauer[1]	none	none	none
G. D. Gurley[1]	oil or gas	none	none
J. W. Crotty[1]	oil or gas	sale	1

NOTES: By terms of the bid, cash due within nine months of receipt of letter of acceptance from the WAA.

[1]Bids ineligible, not accompanied by $100,000 deposit or otherwise incomplete.

SOURCE: War Assets Administration, Proceedings of Reading of the Bids for the Big and Little Big Inch Pipelines, February 10, 1947.

As the winning bidder for the Inch Lines, Texas Eastern became an overnight force in the burgeoning interstate natural gas pipeline business. The Inch Lines attracted the attention of an incredible array of politicians and businessmen, and untold efforts were undoubtedly undertaken by various bidders to ensure a successful outcome. But the rumors and suspicions of unsavory activity surrounding the bid gradually faded before a new reality of the postwar prospects for the expansion of the natural gas industry. Demand was high and prices were low (see table 4.3).

The war years stimulated renewed growth in the natural gas industry. Specifically, increased demand focused attention on the great waste of southwestern natural gas. Bill Murray, a Texas Railroad Commission employee who later became a commissioner, recalled, "I was appalled at the waste." But with pipelines to connect the gas fields with eager customers, gas production would become a profitable venture. The bid for the Inch Lines represented the wartime government's last major involvement in this era of gas industry expansion. The war was a catalyst for the construction and operation of new natural gas pipelines connecting southwestern fields with northeastern markets, but peacetime market

Table 4.3. Average Natural Gas Wellhead Prices
(cents per mcf)

Year	Price
1930	7.6
1935	5.8
1940	4.5
1941	4.9
1942	5.1
1943	5.2
1944	5.1
1945	4.9
1946	5.3
1947	6.0
1948	6.5
1949	6.3
1950	6.5
1951	7.3
1952	7.8
1953	9.2
1954	10.1
1955	10.4
1956	10.8
1957	11.3
1958	11.9
1973	21.6

SOURCE: *Energy Factbook*, Congressional Research Service Library of Congress, 96th Cong., 2d sess. (Washington, DC: GPO, 1980), 490. Also see U.S. Bureau of Mines, *Minerals Yearbook*.

forces would now dictate industry growth. The large metropolitan areas of Philadelphia, New York, and Boston desired gas and could now reasonably expect to receive it. Having won the highly politicized government auction for the Inch Lines, Texas Eastern faced new challenges in the expanding natural gas industry.

5. Pennsylvania Coal and Eminent Domain

AFTER WINNING the bid, Texas Eastern prepared to deliver southwestern natural gas into the Northeast. Coal and railroad interests, integral parts of the manufactured-gas industry, immediately presented strong opposition to Texas Eastern's plans to both acquire and operate the lines. These industries feared the competitive effects of natural gas on their northeastern coal and coal transportation businesses. Coal and railroad companies opposed natural gas primarily through political rather than economic means. A good portion of their political influence resided in Pennsylvania, through which the Inch Lines crossed, and Pennsylvania had adopted an anti–natural gas stance. Although Pittsburgh and other cities in the western part of the state had access to locally produced natural gas, cities in the eastern section, including Philadelphia, the largest energy-consuming metropolitan area in that part of the state near the path of the Inch Lines, had no natural gas. Texas Eastern's founders planned on quickly extending gas sales beyond Appalachia by breaking into the more lucrative metropolitan markets along the Inch Line route such as Philadelphia and New York City.

Texas Eastern targeted Philadelphia, the closest of these cities to Appalachia, to be its first new market. Philadelphia, like other northeastern metropolitan regions, relied upon coal and manufactured gas for residential, commercial, and industrial energy. Much of the coal used in the Northeast originated from Pennsylvania coal mines. Through World War II, coal provided about 50 percent of the nation's energy requirements, and about 41 percent of United States coal was mined in Pennsylvania. This figure included about 30 percent of the nation's bituminous coal, which was used primarily as an industrial fuel, and virtually all the country's anthracite coal, which was used in the East for home heating.[1] The coal industry did not want its traditional markets to be eroded by southwestern natural gas.

EMINENT DOMAIN FOR NATURAL GAS

Perhaps the greatest single obstacle confronting Texas Eastern's efforts to deliver gas into Philadelphia and farther eastward was the lack of a federal eminent domain law for natural gas pipelines. Charles Francis, Texas Eastern's general counsel, recognized this potential problem as early as June 1946 during preparations for the first bid. This complicated Texas Eastern's plans to sell gas into and beyond Pennsylvania; the state had erected similar obstacles to natural gas and actively enforced them to block natural gas expansion into the state. Congress had earlier considered federal eminent domain legislation for gas pipelines, but it approved such a law only for petroleum lines. In 1941, Congress enacted the so-called Cole Bill, which granted eminent domain powers to petroleum pipelines.[2] During the congressional debate of the bill, Congressman William Cole (D-MD) added that although "there was a request" for natural gas to be included in the bill, the committee decided that since the FPC already regulated the natural gas industry, the Cole Bill might not be appropriate. Cole added: "I can very easily see where the future of these lines [the proposed war emergency lines] might be better for natural gas than petroleum if tankers are available to supply the needs of the Atlantic coast."[3] He did not, however, recognize that the FPC could not grant eminent domain powers; if the lines were converted later to natural gas they might very well require a federal eminent domain statute.

Similar problems confronted Texas Eastern on the state level. Both Pennsylvania and its coal industry had long worked hard against the possibility that the Inch Lines would be used for natural gas transportation into the Northeast. When the Defense Plant Corporation originally contracted for the construction of the lines, it acquired the necessary right-of-way permits from the various relevant states for their operation. The state of Pennsylvania, however, carefully agreed to grant right-of-way easements for the transportation only of petroleum or its by-products across the one-hundred plus streams within the state. In addition, the City of Philadelphia and Board of Commerce and Navigation of Delaware County also specifically limited the use of the pipelines to the transportation of oil and petroleum products.[4] Pennsylvania governor James Duff, who was formerly the state's attorney general, resolved any doubt about the effect of the right-of-way law on natural gas when he ruled that natural gas was not a petroleum product. Before the Inch Lines could transport natural gas into the state, or even through it, Pennsylvania would have to permit the modification of the original right-of-way agreement.

The state's official position clearly reinforced the interests of coal miners, railroads, and labor unions. These groups adamantly opposed any modification of Pennsylvania's restrictions against natural gas. D. L. Corgan, secretary of the anthracite committee, stated in November 1946 that "use of the pipelines for gas on a permanent basis would be a disastrous blow to Pennsylvania's economy and would throw thousands out of jobs."[5] In addition, a majority of Pennsylvania's congressional delegation had vainly attempted, but failed, to postpone the federal government's policy decision to sell the Inch Lines to the highest bidder. And on several previous occasions, Pennsylvania congressman Francis Meyers attempted to block the sale of the Inch Lines for natural gas transportation.[6]

Texas Eastern therefore faced a series of operational, regulatory, and financial hurdles before it could fully concentrate on selling gas. The Pennsylvania problem aside, Texas Eastern's managers had to form an efficient organization quickly. The company began its corporate existence in a unique position for a gas pipeline company. Normally, before a new pipeline could be built, it would apply for a certificate from the FPC. But Texas Eastern's high bid gave it provisional ownership of the lines before it acquired certification and financing for the line. The WAA mandated that Texas Eastern arrange for the certificate and financing within nine months from the date it issued a letter of intent to sell the lines. The WAA issued this letter on February 25, 1947, after the U.S. Department of Justice approved the sale.[7] The Justice Department was concerned that Texas Eastern and the United Gas Corporation, the largest gas company in the Southwest, might have some special relationship including United Gas control over Texas Eastern's gas purchase policy. Although United Gas president Norris C. McGowen had once had a financial stake in Poe's early effort to bid on the Inch Lines, he later realized that the pipeline company could best serve United Gas as a purchaser of its own gas production. Thus, Texas Eastern attorney J. Ross Gamble informed the Justice Department that his firm had no special or secret arrangements with United Gas.[8]

In the interim between Texas Eastern's winning bid and April 31, the expiration date of Tennessee Gas's lease, the company prepared to operate the Inch Lines. The company's top priority was recruiting a staff; since the firm began its existence essentially as a promotional venture, it would now have to be transformed from what appeared to be a speculative financial venture into an operating regulated public utility.

Only nine days before the bid, on January 30, 1947, Texas Eastern Corporation had officially been incorporated. At the same meeting, the company announced that it would issue 150,000 shares of stock priced at $1 each to as many as thirty-five associates of the group. Subsequently, George and Herman Brown each purchased 21,375 of these shares. George Brown also lent $67,500 for share purchases, in return for a two-year voting proxy, to several of the other twenty-eight actual investors. This gave George Brown alone control over 59 percent of the company's stock. Adding his brother's shares, the Brown brothers controlled 74 percent of the firm's original stock. Chairman of the board George Brown was, therefore, firmly in control of the company.

At the time of bid, E. Holley Poe was president of Texas Eastern. He was indisputably the group's founding member and the man who guided the project through to the winning bid. His management skill, particularly in such a large and then unorganized firm, however, was questionable. The company's organizers, recognizing this, had previously agreed that if the bid was successful, the highly experienced and respected Reginald Hargrove would become president, replacing Poe. In fact, there was an effort to squeeze Poe out of the whole deal, but Poe's friend and fellow founder, Everette DeGolyer, threatened to quit the venture if Poe was not to be included.[9] Poe had other problems as well. His founder's share of stock, 10,875 shares, diminished considerably after he turned over one-third of it to Ashland Oil as compensation for an interest in the venture taken much earlier by J. Howard Marshall of Ashland Refining Company.[10]

Reginald Hargrove, however, was a strong manager who could lead Texas Eastern into the competitive natural gas pipeline industry. Quickly assuming his new role, Hargrove brought with him numerous managers and engineers from United Gas to run the new company. N. C. McGowen did not complain as Hargrove and others including his general counsel, George Naff, and assistant chief engineer, Baxter Goodrich, left his company for Texas Eastern. In return, Texas Eastern promised United Gas a guaranteed market for its gas production and supply. Other new executives at Texas Eastern included Herbert Frensley, secretary and treasurer, who was a trusted Brown & Root official, and Orville Carpenter, comptroller, who was an associate of the Browns and previously the executive director of the Texas Unemployment Compensation Commission, as well as associates of George and Herman Brown. Texas Eastern's other personnel came largely from the original field crew of the War Emergency Pipelines.[11]

As Hargrove worked to put together an operating group, August Belmont of Dillon, Read, began arranging the necessary $120 million in financing for the venture. Dillon, Read was no stranger to financing gas ventures. According to Belmont, the firm was the first to have "gas bonds approved as legitimate investments for life insurance companies."[12] Over the course of the following nine months, Belmont arranged for the financing through northeastern life insurance companies. The remainder of the purchase price of the lines was intended to come from a stock offering to be made after the FPC issued the certificate.

Generally, the process of organizing Texas Eastern proceeded smoothly, and the company achieved its first goal when the FPC issued it a temporary certificate on March 21. In addition, the gas firm began to contract for gas sales. As Tennessee Gas was still selling gas through the leased Inch Lines in large Appalachian markets, Texas Eastern representatives did not have to look far for customers; in fact, the FPC required Texas Eastern to continue selling gas to the same Appalachian distribution companies. Texas Eastern also applied for the required permanent certificate of public convenience and necessity, and the FPC scheduled hearings on the application to begin in July 1947.

As Texas Eastern organized and prepared to take possession of the Inch Lines on May 1, the Pennsylvania right-of-way problem emerged as the group's most formidable problem. Poe, however, had always been optimistic. Within days of the winning bid and in response to a newspaper reporter's question about Pennsylvania's seemingly intransigent position against natural gas transportation through the Inch Lines, Poe had remarked that "there has been interstate commerce in this country ever since the Constitution was adopted," and he did not expect Pennsylvania to actually block the interstate transmission of natural gas through the state.[13] Poe's statement reflected the confidence of the new Texas Eastern. George Brown was apparently less optimistic when he said, "The speed with which we can reach our maximum planned rate of delivery depends upon how soon right-of-way possession, deeds, and other legal matters can be completed. This may take from eight to 15 months."[14]

In fact, Texas Eastern general counsel Charles Francis had been at work on the right-of-way problem for over a year. Soon after Texas Eastern's founders first organized in early 1946, Francis recognized the potential problem that the lack of a federal eminent domain law could impose on the group's efforts. He identified persons associated with the group's members who might assist Texas Eastern's lobbying efforts. One of them was Frank Andrews, Reginald Hargrove's cousin-in-law, who had connections with both financial and political interests. In particular, Andrews was president of the Hotel New Yorker, which was owned by Manufacturers Trust Company—the firm that later provided Texas Eastern with invaluable loans and financial assistance. Perhaps even more important, Andrews had introduced George Allen to the group the previous summer. Allen, a close adviser of President Truman and director of the RFC, was one of Washington's most celebrated lobbyists. Francis reported to his superior, Judge Elkins:

> Mr. Allen . . . is coming here Monday for a conference with Hargrove, Poe, and Andrews. I have generally outlined the form of our bid and the suggested legislation which we lack as a condition to our bid, to wit: a congressional authorization for exercising the power of eminent domain and a permit from the Federal Power Commission to transport gas interstate. . . . I plan on having a talk with Mr. Allen in Washington next week.[15]

Francis's efforts to obtain a federal eminent domain ruling continued throughout the next year. It is not clear exactly what role Allen played, although one knowledgeable source recalled that Allen was prominently involved in the venture.[16]

By mid-February 1947, Governor Duff appeared to be wavering in his anti–natural gas stance. He told a news reporter that Pennsylvania was virtually powerless to prohibit the transmission of natural gas through the state. Duff was appropriately sensitive to energy needs in his state, and he did not prohibit those areas of the state already using natural gas from receiving new supplies. Senator Myers (D-PA) was significantly more optimistic about the possibility of keeping natural gas out of Pennsylvania when he noted that the FPC would "never permit the transmission of natural gas to the eastern seaboard."[17]

Texas Eastern's interest in the Pennsylvania gas market was tempered by an assortment of other obstacles. One was the reluctance of the state of Louisiana to rescind its ban on the exportation of natural gas from its borders. Another more immediate one was the need to defend itself against an assortment of attacks by competing bidders. Gardiner Symonds's Tennessee Gas led the way in questioning the validity of Texas Eastern's bid. "I think Gardiner was very upset," recalled a Texas Eastern official, "that we got a bid that was higher than his, and he was god damned if we were going to go and get that away from him."[18] Symonds's aggressiveness was once again evident soon after Texas Eastern won the Inch Lines. General Littlejohn had returned the $100,000 deposit checks to each losing bidder. Tennessee Gas, though, sent its deposit back to the WAA stating that it was returning the deposit in order to keep its bid open in case Texas Eastern's bid proved inadequate. The WAA soon returned Tennessee Gas's deposit again, and this time the company seemed to accept defeat.

Coal and railroad interests also officially protested to the FPC about Texas Eastern's planned use of the Inch Lines to transport natural gas. Attorney Tom J. McGrath, representing eight coal, railroad, and labor organizations including the National Coal Association, Eastern Gas and Fuel Associates (a coal and manufactured-gas company), the United Mine Workers of America, the Chesapeake & Ohio Railway, and several railroad workers labor unions, urged the agency not to grant Texas Eastern a temporary certificate to operate the Inch Lines. McGrath charged that the FPC lacked jurisdiction to grant Texas Eastern a temporary certificate without prior notice and hearing. The FPC, however, asserted its right to do so and worked toward granting such a temporary certificate to Texas Eastern so that it could begin operating the pipelines on May 1, 1947. To this end, the WAA informed Tennessee Gas on March 8 that it would not be renewing its lease of the Inch Lines to Tennessee and that "it is hoped that [you] will cooperate in the delivery of these lines to Texas Eastern Corporation."[19]

Tennessee Gas found McGrath's efforts to be worthy of its own support, at least in this one very specific case. In letters to the FPC, Tennessee cited McGrath's argument that the FPC lacked jurisdiction to grant a temporary certificate to a new natural gas company without a hearing.[20] Texas Eastern responded that it would simply continue to operate the Inch Lines under the same emergency conditions as Tennessee had operated them. In this sense, then, the Inch Lines were not a brand new company but an existing service.

Tennessee Gas tried a new tactic to thwart Texas Eastern. In an attempt to resolve potential competition outside the regulatory arena, representatives of Texas Eastern and Tennessee Gas met on March 3. Gardiner

Symonds and three others represented Tennessee Gas while R. H. Hargrove and Charles Francis represented Texas Eastern. They discussed their understanding of potential conflicts. Hargrove stated that Texas Eastern could not finance its purchase of the Inch Lines without first signing long-term contracts with customers in the northeastern states. This meant selling gas into and east of Appalachia. Hargrove suggested that Texas Eastern might sell 250 mmcf/d in an eastern area such as Brooklyn and 200 mmcf/d in the Pittsburgh area, and he hoped that this would not disturb Tennessee Gas. A representative of Tennessee Gas responded that his company would oppose through all legal means any move by Texas Eastern into its own current service area. Hargrove replied that Texas Eastern would then be forced to oppose any expansion of Tennessee Gas's facilities.[21]

Discussion continued and the participants made several different proposals aimed at dividing the northeastern markets between their companies, but the meeting ended with no clear resolution to share markets. Tennessee Gas and Texas Eastern gave up the idea of privately negotiating the division of particular markets between themselves. From this point forward, the two companies competed for market share primarily through the regulatory process and conducted the bulk of their negotiations for market area expansion within the confines of the FPC certification process to be played out in upcoming hearings.

While Tennessee Gas and Texas Eastern squabbled over their rights to particular market areas, the severe winter of 1947 emphasized the need for increased fuel supplies for the Northeast. During the week of February 10, Pittsburgh newspapers reported that diminishing supplies of industrial gas resulted in 50,000 temporary local layoffs. On February 12, Tennessee Gas's pipeline suffered a break, further aggravating the shortage in the Pittsburgh area.[22] The great demand for natural gas in Appalachia and the Northeast was in fact much larger than any single pipeline company could supply. A new prospective competitor entered the fray when the Memphis Natural Gas Company indicated its interest in building a 24-inch line connecting Texas and the Pittsburgh area for $63 million. But this proposal never progressed beyond the planning stages.

The controversy surrounding the ownership of the Inch Lines was lost to many northeasterners, who cared only about access to fuel to run their businesses or heat their schools and homes. The WAA received praise from editorial writers and others glad to know that the Inch Lines would finally and permanently transport natural gas to the fuel-short Northeast.[23] Conversely, coal interests continued to oppose the introduction of natural gas, particularly for industrial use, in the region. Tom McGrath remained on the offensive, proposing that natural gas should be restricted to cooking and heating. McGrath remained concerned that natural gas would be used to displace the industrial coal-burning market. He opposed natural gas use for steam generation and endorsed its use only in current market areas. McGrath noted that natural gas sold through pipelines had displaced 50 million tons of coal a year involving 32,000 worker days and about half as many railroad worker days.

After the FPC scheduled hearings on Texas Eastern's permanent certificate application, the agency notified all potential interested parties that it would be holding public hearings in consideration of Texas Eastern's application. To Texas Eastern's consternation, the Pennsylvania Public Utility Commission responded immediately on May 6 with a petition to intervene in the hearings. In a cover letter, J. A. Walter, acting secretary of the Pennsylvania PUC, stated that "the Pennsylvania Commission is of the opinion that the total deliverable capacity of the transmission lines comprehended within the above application should be restricted to Western Pennsylvania and that delivery therefrom should not extend eastwardly beyond the Pittsburgh area."[24] The PUC's position was an unwanted potential obstacle to Texas Eastern because the PUC had authority to regulate the service, rates, and charges of natural gas operations in the state.

George Brown responded to the PUC's action in a speech given to the American Society of Civil Engineers in Houston in late May. Brown argued that Pennsylvania had created "a barrier against the use of the Big Inch and Little Big Inch lines in gas transmission." Brown suggested that by being forced to compete for markets with natural gas, the coal interests might be motivated to bring improvements to its industry. "Probably some day," Brown said sarcastically, "the coal industry will praise us for stimulating it to more efficiency."[25]

The upcoming FPC hearings for Texas Eastern's certificate of public convenience and necessity promised to become a significant battleground on which Texas Eastern would seek permission to market natural gas to eastern markets embedded in manufactured-gas territory. Coal and railroad interests could be expected to marshal an all-out effort to oppose Texas Eastern, which would need to present a strong case that the Northeast desperately needed natural gas. Thus, preparations to prove this need before the FPC went forward at Texas Eastern, various agencies, and companies wanting gas.

But as Texas Eastern prepared to pursue its interests before a federal regulatory agency, the coal industry's strong political influence in Pennsylvania threatened its progress at the state level. Pennsylvania's restrictive right-of-way laws blocked any attempt to build new gas pipelines through the state or market gas to the existing manufactured-gas utilities in eastern Pennsylvania. Texas Eastern's promoters realized that their company's success rested upon access via the Inch Lines to the large metropolitan markets east of Appalachia. Without access to the large utilities and industrial consumers in Philadelphia and New York, Texas Eastern's gas sales would fall well below levels necessary to generate funds to pay for operations and service the company debt.

As this would place the company in financial jeopardy, several of the insurance companies with which Texas Eastern negotiated for financing indicated to Charles Francis that they might withdraw from participation, in the absence of federal powers of eminent domain. Francis responded by lobbying for congressional passage of a federal eminent domain statute for natural gas pipelines.[26] He worked with several politicians to prepare appropriate bills from Texas Eastern's perspective. In early April, Senator Edward H. Moore (R-OK) introduced such a bill, S. 1028. His cointroducers were Senators Thomas Connally (D-TX), O'Mahoney (D-WY), and Arthur Stewart (D-TN). In the House, Representative Schwabe (OK) introduced a similar bill, H.R. 2956. According to Francis, he and Vinson, Elkins, Weems & Francis attorney David Searles, among others, drafted the bill.[27]

The eminent domain bill was only one of several highly controversial amendments proposed in 1947 to the Natural Gas Act of 1938. Perhaps for this reason, the bill received less attention than it might have if it had been proposed alone. During April and May, the Committee on Interstate and Foreign Commerce held hearings on five separate proposed amendments.[28] These included the Rizley amendment (H.R. 2185), which was designed to limit the power of the FPC and explicitly exempt gathering, production, and local distribution from FPC regulation.[29] The Dolliver bill (H.R. 2569), designed to increase the powers of the FPC, was supported by coal and railroad companies that believed a more powerful FPC would inevitably place more restrictions on natural gas to the benefit of coal and coal-transporting railroads.

Some opposition to the eminent domain bill did filter through the acrimonious debate on the Rizley and Dolliver bills. Tom McGrath, the forever present elder statesman of the anti–natural gas lobby, represented at these hearings the National Coal Association, consisting of 85 percent of the nation's bituminous coal production; Eastern Gas & Fuel Associates, a large bituminous coal producer; and the Chesapeake & Ohio Railway Company, the nation's largest transporter of bituminous coal. McGrath indicated that he did not understand at the opening of the hearing that the eminent domain bill would be discussed. Being unprepared to mount an effective opposition, McGrath stated briefly that the issue needed much further discussion, especially in regard to the bill's possible violation of state's rights.[30]

In the press, the *Oil and Gas Journal* remarked that although no natural gas pipelines had yet had difficulty in attaining rights-of-way, some proposed emergency oil pipelines had experienced such difficulties during the war years and these restrictions were resolved by the Cole Bill. The article noted that if Texas Eastern received an FPC certificate to transport gas through Pennsylvania and the eminent domain bill also passed, "Texas Eastern would have power to compel the state to give it an easement for gas also."[31] With federal powers of eminent domain, no state government could permanently stand in the company's way.

Francis also discussed the necessity of the eminent domain bill with August Belmont. Francis reported that "it is highly important for us to see that the present Congress adopts the Eminent Domain Statute, which I have been pushing. . . . I think a federal statute is highly important, if not almost essential, in straightening out our right of way difficulties."[32] The eminent domain bill moved through Congress during June without the expected opposition.

Meanwhile, Texas Eastern prepared to take possession of the Inch Lines, which it did on May 1, 1947, from Tennessee Gas. Although Gardiner Symonds remained angry over losing the bid, he had no choice but to relinquish the lines, and Texas Eastern agreed to pay Tennessee Gas approximately $240,000 to cover Tennessee Gas's expenses in its efforts to transmit natural gas through the petroleum lines. The transfer took place in a conference call involving Texas Eastern, Tennessee Gas, and War Assets Administration personnel. The crews actually operating the line did not change and many continued working for Texas Eastern.

As of May 1, Texas Eastern also instituted its first gas purchase policy. It announced it would pay $0.08 per mcf on a twenty-year contract and $0.065 on a five-year contract. Sellers would have to prove a 150 percent reserve for the life of the contract.

The company also commenced negotiations for gas sales contracts with the Philadelphia Electric Company and the Philadelphia Gas Works

Company. Although Reginald Hargrove had told Gardiner Symonds during their conference in early March that Texas Eastern intended on contracting for gas sales to utilities in Pittsburgh, Texas Eastern now publicly indicated that Philadelphia would be its first market east of Appalachia. Philadelphia was on the outskirts of Appalachia and close to the paths of the Big Inch and the Little Big Inch. Philadelphia was the third-largest city in the United States with a strong manufacturing base. Wholly dependent on manufactured gas, it offered an obvious opportunity for Texas Eastern.

PHILADELPHIA GAS DISTRIBUTORS

The city had two major utility systems, the Philadelphia Electric Company and the Philadelphia Gas Works Company. Although both sold manufactured gas in the Philadelphia area, the manufactured fuel became increasingly more expensive during and immediately after World War II, and both utilities were eager to contract natural gas supplies. Both utilities intended to use new natural gas supplies to produce mixed gas, since Texas Eastern alone could not provide enough natural gas to justify an immediate and complete conversion to the new fuel.

The Philadelphia Gas Works Company was a municipal gas works operating under a franchise agreement with the city since 1898. The company operated under the supervision of the Philadelphia Gas Commission, a five-man body, one of whom was an official of the company; the others were city officials or public designees. The company manufactured carbureted water gas and coal gas in two plants, purchased certain by-product coke oven gas from the Philadelphia Coke Company, and distributed and sold this gas to some 500,000 customers in Philadelphia. The company also operated the Northern Liberties Gas Company in a section of Philadelphia with 4,000 customers; these two companies were the sole suppliers of gas service in the city limits of Philadelphia. The Gas Works distributed mixed gas, largely made up of carbureted water gas, with a 530-Btu content to its customers.[33]

The Philadelphia Electric Company sold both electricity and gas in the Philadelphia metropolitan area. It provided electricity to customers in the city and surrounding areas and manufactured gas to areas outside the city limits. The Philadelphia Electric Company had been engaged in the manufactured-gas business since the early twentieth century. During the late 1920s, the company's gas sales to suburban customers began to decline as electricity captured the lighting market, but consumers continued to use gas for both cooking and heating. Demand for gas heating slowly increased, and the utility enlarged its small manufactured-gas production capacity by reforming the manufactured gas with refinery oil. By 1929 the company increased its gas sales to a total of 868 home heating customers. Its customer base increased gradually from 2,149 in 1935 to 3,085 by 1940, but this represented a very slow growth rate compared to the increase in home electricity consumption. The financial constraints of the depression had limited the growth of the manufactured-gas business. Beginning in 1939, however, manufactured-gas use grew, due primarily to the increased energy demands of the war-related industries as wartime energy-use restrictions reduced the ability of utility companies to expand service for nonessential customers. Industrial gas sales, including those for war production, multiplied fivefold while total gas sales increased from 44 to 74 mcf/d.[34] However, Philadelphia Electric's overall position regarding gas use was typical of other utilities in the area. The company's management felt ambivalent about its relatively small and stagnant manufactured-gas business, which had barely grown in the previous decade. Systemwide gas sales of the previous ten years increased by only 6 percent, while electric sales increased more than 85 percent.[35]

In August 1944, the War Production Board lifted its restrictions on Philadelphia Electric and other gas distribution companies. The restrictions had required the distributors to provide service only to those customers who were considered hardship cases by the War Production Board. Lifting of restrictions was followed by a surge in gas-heating installations. In 1945, prospective customers swamped the company with orders for gas service, straining its production and transmission capabilities. The company's Chester gas plants underwent a renewed expansion to increase capacity by 12 percent and new mains were laid as well.[36]

After the war, gas heating became widely popular. In 1946, Philadelphia Electric alone had 9,000 orders for gas installations. But during that year, the company could not come close to meeting the demand. In fact, it installed only 1,721 new connections out of its total 9,000 orders. The primary limiting factor was the lack of available gas supply. The Philadelphia Electric Company depended upon manufactured gas, and substantially increasing the company's manufactured-gas output required a significant capital investment in its gas-manufacturing facilities. Even the recent 10 mmcf/d increase in manufacturing capacity at the Chester gas plant did not begin to make up for the new demand.

The Philadelphia utilities could easily present a strong customer demand for natural gas service. Texas Eastern's attorneys worked closely with representatives of the utilities to prepare and coordinate their upcoming FPC testimony regarding the demand for gas. Also, by the end of the month, Texas Eastern and the Philadelphia Electric Company and the Philadelphia Gas Works agreed to the general form for proposed gas sales contracts. Both the City of Philadelphia and the Philadelphia Gas Commission then petitioned the FPC to intervene in Texas Eastern's upcoming hearings in support of the sale of natural gas in Philadelphia.

The city's petition stated that Philadelphia owned and controlled the Philadelphia Gas Works under an operating agreement. The city valued the Gas Works at "in excess of $100,000,000 and [it] is engaged in furnishing gas for domestic heating, cooking, lighting and manufacturing provided to inhabitants of a city of more than two million people . . . [and] provides service to approximately 500,000 gas consumers in the city of Philadelphia, sales currently running 30,000,000,000 cubic feet of gas per year." The petition also maintained that the city had "a constantly increasing demand for gas resulting from the construction of new and additional housing."[37] The Philadelphia Gas Works proposed to use natural gas to offset the present need to construct additional manufactured-gas facilities, "the financing of which under present conditions being doubtful." The gas companies realized that the tremendous expense of constructing additional manufactured-gas plants would raise the price consumers would pay for gas. But simply purchasing natural gas from a pipeline cost only the price of the gas, one-time charges for building a line to connect with Texas Eastern, and the cost of adjusting customer appliances.

The city's petition also noted that the Philadelphia Gas Works Company and the City of Philadelphia had negotiated a gas sales agreement with Texas Eastern. But before this sale could go forward, the Philadelphia Gas Commission had to recommend the transaction to the city council, which would then have to pass an ordinance approving the sale. The petition stated that the city council was currently in the process of approving an ordinance to allow Texas Eastern to sell natural gas to the Gas Works. Philadelphia's clear support of natural gas purchases by its utilities provided Texas Eastern with a substantial case in favor of its overall operating plans to be presented at the upcoming FPC hearings.

The Philadelphia Gas Commission likewise petitioned the FPC to intervene in the hearings and support Texas Eastern's gas sales agreement with the Philadelphia Gas Works. This petition repeated that of the City of Philadelphia and noted that on June 24, 1947, it approved by resolution the sale of natural gas by Texas Eastern to the Philadelphia Gas Works.

TEXAS EASTERN'S CERTIFICATE HEARINGS

The FPC hearings for Texas Eastern's certificate began on July 7, 1947, attracting a panoply of support and opposition from forty-eight intervenors.[38] The majority of the intervenors were supporters of Texas Eastern's certificate, most of which had contracted to receive gas from Texas Eastern, or hoped to do so; these companies represented a demand for natural gas that Texas Eastern alone could not fill. These firms included the New York State Natural Gas Commission, East Ohio Gas Company, Equitable Gas Company, Philadelphia Electric Company, and the Philadelphia Gas Works. The opposing intervenors included the usual assortment of coal, railroad, and labor interests as well as the competing natural gas companies of Tennessee Gas, Panhandle Eastern, and the unbuilt Transcontinental Gas.

On the second day of the hearings, the Commonwealth of Pennsylvania intervened in the hearings and indicated an important change in its position. Deputy Attorney General John C. Phillips spoke in favor of natural gas importation, noting that a shortage of gas now existed in Philadelphia as well as Appalachia. The commonwealth desired to assure the state that it would have an adequate supply of the fuel, and it now supported Texas Eastern's sales agreements with the two Philadelphia utilities. This position was in sharp contrast to the commonwealth's vigorous opposition to natural gas imports voiced less than a year earlier. The continuing gas shortage in the state had helped to convince officials to temper their earlier defense of coal and acquire a more positive view of the importation of natural gas. State officials had come to appreciate that new and developing industries might be interested in locating in areas of the state with access to natural gas.[39]

Legal counsel represented Texas Eastern as well as each intervenor. These lawyers had the power to question and cross-examine each other during the course of the hearings. David T. Searles of Vinson, Elkins, Weems & Francis along with J. Ross Gamble, R. Clyde Hargrove, Charles I. Francis, and Jack E. Head represented Texas Eastern.[40] In many respects the hearings developed into a battle between David Searles and the bespectacled long-time coal grandee Tom J. McGrath. During these hearings, McGrath represented on his own or in conjunction with other attorneys the Anthracite Institute, Baltimore and Ohio Railroad (and other eastern railroads), the Chesapeake & Ohio Railway Company, Eastern Gas and Fuel Associates, Eastern States Retail Solid Fuel Conference, the National Coal Association, and the United Mine Workers.

Searles, as Texas Eastern's lead attorney in the hearings, explained in his opening statement to the FPC the basis of the company's application. Texas Eastern's case included substantial detail about marketing, engineering, accounting, gas supply, finances, and management. Much of the hearing was a tedious and perfunctory exercise in presenting masses of data designed to prove to the FPC's satisfaction that Texas Eastern was a competent organization staffed by experienced personnel—drawn from the well-established United Gas Corporation and the War Emergency Pipeline—with adequate supply for its proposed markets.

Searles raised two closely related issues that were the heart of Texas Eastern's problems: its interest in contracting gas sales to Philadelphia and acquiring federal rights-of-way for its natural gas pipeline. Searles introduced these issues by testifying that in addition to a projected delivery of 340 mmcf/d of natural gas to the Appalachian region, the company planned to sell 80 mmcf/d to two utility companies in Philadelphia, leaving an excess of 13 mmcf/d for sales flexibility. Texas Eastern strongly desired to break into the Philadelphia market, which was among the largest on the East Coast. The company's gas sales contracts with the Philadelphia Electric Company and Philadelphia Gas Works Company were more arguments in favor of natural gas sales into eastern Pennsylvania.

Texas Eastern's most powerful and not so secret weapon, the eminent domain bill introduced to the House by Representative Schwabe in early April, was gradually making its way from committee to the House floor. Finally, on July 25, 1947, the statute became an amendment to the Natural Gas Act of 1938 when President Truman affixed his signature to it. It is not entirely clear why only limited opposition to this bill was presented to Congress, but it is evident that the strong demand for natural gas in the Northeast persuaded authorities to support rather than oppose the eminent domain amendment. For the first time, natural gas companies had the support of federal eminent domain powers to counter any obstacles that states or railroads might try to impose on them.

With this issue finally removed from the table, Texas Eastern could at last move on beyond Appalachia into Philadelphia and the Northeast.

This was not simply a matter of an expansive company wanting to increase its sales and profits. The company had real fears that delays in shipping gas to markets along the Inch Lines might have ramifications for its rate base. It was not clear if the FPC would allow the company to include in its rate base the costs of any portions of the lines not put to use in transporting gas, such as those portions extending east beyond Appalachia to New York. Texas Eastern required no answer to such a question; from the company's point of view, it was past time to move on to the major cities of the East.

Stripped of the right-of-way issue, intervenors from the coal industry introduced a new issue into the FPC hearings. They argued that Texas Eastern should be forced to limit its sales to the fuel-short Appalachian region. By convincing the FPC of the pressing need there for all the gas available to Texas Eastern, they sought to limit the company's expansion into the much larger traditional coal markets to the east. These intervenors focused on the company's plans to extend service to Philadelphia, objecting to this prospective "diversion" of gas from the Appalachian region to areas not affected by energy shortages where natural gas would compete with coal for large residential and industrial markets.

McGrath pressed Reginald Hargrove on this point during the hearings. McGrath asked Hargrove, "If the Commission should see fit to condition the certificate in this case so as to require deliveries to be made in the Appalachian area or west thereof, do I understand that you would not accept such a certificate?" Hargrove responded smartly: "I haven't stated that, Mr. McGrath. What I did say was that if such a condition was imposed in a certificate, it would make a reconsideration of the whole matter necessary, and I don't know that it would be feasible in any event, but I do know that in the short time remaining between now and November 25th it would be practically an impossible undertaking to recast the entire project along those lines."[41] McGrath then asked Hargrove if Texas Eastern would continue to operate the pipelines under the terms of its lease through May 1, 1948, if it did not receive a certificate for sales east of Appalachia. Hargrove responded that although the WAA had authority to extend the November 25 deadline for the complete financing of the Inch Lines, "I do not think that if the certificate were delayed to the point that we could not complete our deal by November 25, some of the same interests that are urging there is no need for the certificate to be issued promptly would urge the Administrator not to extend it, and I don't want to be put in that dilemma if I can avoid it."[42]

Although Texas Eastern's underwriters and the company itself desired entry into the markets for natural gas east of Appalachia, a number of intervenors in Appalachia and westward testified that they should also receive some of Texas Eastern's supply. Some communities desired natural gas so fervently that they sent representatives to the hearing to request that the FPC allocate gas to them from other pipelines not even involved in the hearings. Bloomington, Illinois, was one such community. Although located 200 miles from the Inch Lines, Bloomington requested a natural gas supply from Panhandle Eastern Pipe Line Company to replace its existing manufactured-gas service. The commission responded by requesting the city to come back when Panhandle Eastern's sales were under consideration. It did, however, grant requests from several intervenors for small amounts of gas.

The FPC noted that the gas shortage in Appalachia and the surrounding area would probably continue into the next winter. A portion of this area was supplied by Panhandle Eastern Pipe Line, which estimated that distributors in its service area required 518 mmcf/d. Since Panhandle Eastern's own capacity was 425 mmcf/d, a significant shortage existed. The FPC required Texas Eastern to supply 20 mmcf/d in displacement gas to the Panhandle area through April 30, 1948. The commission also reserved the right to require Texas Eastern to do the same the following winter if the shortage continued.

Despite difficulties in meeting existing demand, Panhandle Eastern— whose primary markets were in Illinois, Indiana, and Michigan— intervened against Texas Eastern's gas sales in the Appalachian area for fear of losing some of its own markets to the new pipeline company. Tennessee Gas Transmission Company presented the same objection to Texas Eastern's certificate application. Noting, however, that "the positions of the Tennessee Gas Transmission Company and Panhandle Eastern Pipe Line Company were not made entirely clear on the record," the commission stated that "we are not persuaded that the certificate herein authorized constitutes any impairment of the markets of these intervenors."[43] Such seemingly token opposition posed no problem to a company whose system extended well beyond the systems of both Tennessee Gas and Panhandle Eastern.

Furthermore, Pennsylvania's once adamant opposition to the Inch Lines disintegrated entirely. On Thursday, September 23, Texas Eastern attorney J. Ross Gamble telegraphed Francis: "John Phillips advises that Governor has executed the right of way permits." The next day, Governor

Duff announced that he would fully support the importation of natural gas into Philadelphia. Duff had previously stated that his decision on supporting the introduction of natural gas into Philadelphia would depend on the city's interest in receiving natural gas. After receiving hundreds of letters of support for natural gas, Duff directed his representatives to advise the FPC that the state " 'vigorously urges' approval of transmission of the fuel to the city."[44] Hudson W. Reed, president of the Gas Works, echoed the sentiment of the people living in Philadelphia who desired natural gas. Reed reported that "over the past fifteen years there has been a growing demand on the part of both old and new owners of automatic heat, and builders have found that it was extremely difficult, if not impossible, to sell homes with coal heating systems. As a result, they first used oil and subsequently gas in order to provide the automatic heat that their new home purchases demand."[45]

The Gas Works noted that the refinery oil it used in the manufactured-gas process was becoming scarce and much more expensive. Citing a $609,000 increase in its oil costs through the first nine months of 1947, the Gas Works stated that without natural gas to replace the oil in the manufactured-gas process, substantial rate increases would be inevitable. Also, Frederic D. Garman, president of the Philadelphia city council, telegraphed the governor: "The program of building new homes for veterans and others in Philadelphia, as well as our industry, will be adversely affected if Philadelphia does not receive the natural gas it has contracted for. We are counting on your help."[46]

With Philadelphia's demonstration of a clear need and demand for natural gas, Tom McGrath tried a new strategy. He argued that Texas Eastern had not proved that it had contracted for the necessary supply of natural gas to meet its sales contracts. McGrath also tried to show that Dillon, Reed & Company's commitment to finance the deal was dependent upon Texas Eastern's ability to contract for reserves.[47] However, McGrath's new arguments proved unconvincing. The Gulf Coast region had tremendous quantities of gas and Texas Eastern had presented its supply contracts and financing ability convincingly.

On October 10, 1947, the FPC granted a permanent certificate to Texas Eastern. The FPC order did not limit the market area in which the company could sell natural gas. The FPC stated that there was no "logical justification for providing unlimited gas service in the so-called Appalachian area and at the same time denying service to the Eastern Pennsylvania area."[48] However, the Pennsylvania service was not scheduled to begin until October 1, 1948, when various pipelines connecting Texas Eastern's pipeline to the utilities were scheduled for completion. In the interim, the commission allowed Texas Eastern to use the two 20-inch sections of the Big Inch and Little Big Inch extending from Skippack Junction to Linden, New Jersey, for gas storage of approximately 60 mmcf of gas, which was to be reserved for emergency use in Philadelphia. Anticipating such a ruling, the Philadelphia city council had previously approved an allocation of $7.5 million for improvements and a pipeline extension from the Gas Works to meet Texas Eastern's pipeline system.

Confronting a certified Texas Eastern, some coal men, now acting more like good businessmen, moderated their stance against natural gas. James Williams, president of the Pennsylvania Retail Coal Merchants Association, explained that "there has been an erroneous impression that we are strictly coal dealers. In reality we are fuel dealers, dealing in coal, coke, oil and when natural gas comes along, we will deal in that too."[49] However, other coal industry representatives continued their fight against the introduction of natural gas and stated their intention to meet with the governor about their position.

FINANCING THE WINNING BID

Texas Eastern faced one more hurdle: financing. To complete its purchase of the pipelines, Texas Eastern had to pay the government $143,127,000 by November 25. Under the direction of August Belmont, Texas Eastern planned to sell $120 million in bonds and approximately $32 million in stock. The stock was to be sold to investors at $9.50 per share. This represented a potential bonanza for the original purchasers of Texas Eastern's first 150,000-share issue. These founding stockholders purchased shares for $1.00 each during early 1947 before the company had submitted its bid to the WAA. After the successful bid, the company split each of the founding stockholders' shares seven times, which meant their original shares were effectively purchased for about $0.14 each while on the open market Texas Eastern shares would be sold for $9.50 each.

Although the SEC unanimously approved Texas Eastern's stock issue plan, several newspapers reported the deal and Texas Eastern suffered its first dose of negative publicity. One headline read, "Texas Eastern Transmission Stockholders Stand to Make $9,825,000 for $150,000 [investment]." The controversy increased when one founding stockholder, J. Ross Gamble, stated his intention to sell his original shares costing $2,500 for $166,250. The gist of the bad press was that twenty-eight individuals had purchased a $143 million pipeline with only $150,000 and reaped enormous profits at the same time.

To minimize the bad press, August Belmont and C. Douglas Dillon held a series of press conferences to answer reporters' questions about the transaction. Belmont spent the rest of the week discussing the financing with reporters and editors from *Time*, *Newsweek*, the *New York Times*, and other journals. This highly public debate over high finance was later repeated with more subtlety in two opposing articles in the prestigious *Harvard Business Review*.[50]

Belmont's close work with the group had earned him an invitation earlier in May to join the Texas Eastern board of directors, which he did. Without realizing it, Belmont had joined the board in violation of the Natural Gas Act, which prohibited persons profiting from the sale of securities to be officers or directors of the company issuing the security. Apparently, neither Dillon, Read nor its law firm of Shearman & Sterling caught the glaring mistake for weeks. Just prior to completing the financing for Texas Eastern, recalled Belmont, a Dillon, Read executive requested Shearman & Sterling to review again the legality of Belmont's relationship with Texas Eastern. This time, the provision of the act prohibiting the relationship was discovered and Belmont resigned immediately after being informed; once again, the entire Inch Lines deal nearly fell through. Later, though, a lawsuit was filed against Belmont and Dillon, Read for being in violation of the Natural Gas Act. As it began to appear that this slip might reopen the entire Inch Lines transaction to federal inspection, Fred Eaton, managing partner of Shearman & Sterling, explained personally both to the FPC and the U.S. attorney general that his firm had given Belmont the "go ahead" to join and that Belmont had left the board "within one minute" after the mistake was discovered.[51]

The public controversy over intrigue, political entrepreneurship, and dramatic profits gradually blew over, leaving Texas Eastern positioned to close its purchase of the Inch Lines on November 14, 1947. This proved to be a complex closing with 150 representatives of the various parties involved in the transaction. In the end, George Brown, Reginald Hargrove, and Harvey Gibson of Manufacturers Trust gave a check in the amount of $143,127,000 to Robert Littlejohn of the WAA. When General Robert Littlejohn retired one week later as WAA administrator, he showed President Truman Texas Eastern's $143,127,000 check. Truman, who had

Table 5.1. Texas Eastern Transmission Corporation Operations, 1947–1954

Year	Gas Sold and Transported (bcf)	Operating Revenues ($ million)	Miles of Pipeline	Personnel	Reserves (tcf)
1947	39	9	2,733	386	2
1948	118	32	3,028	838	3
1949	162	45	3,203	1,200	3
1950	290	74	3,395	1,290	4
1951	339	84	3,442	1,381	4
1952	371	94	4,217	1,635	7
1953	418	136	4,807	1,755	7
1954	430	145	5,158	1,819	7

SOURCES: Texas Eastern Transmission Corporation, *Annual Report* (various years), and Moody's Public Utility Manual.

made the entire venture possible by signing the federal eminent domain bill, complimented Littlejohn on his success at the WAA and remarked that Littlejohn had shown him "the biggest check I have ever seen." From vital conduits of petroleum for the Allied war machine to war surplus property, the Inch Lines were now beginning a third life as one of the nation's most important natural gas pipelines (see table 5.1).

SELLING NATURAL GAS TO PHILADELPHIA

Although Texas Eastern successfully purchased the Inch Lines and acquired permission to sell gas in Philadelphia, the Philadelphia problem now took a different form. The severe winter of 1947–48 prompted Texas Eastern's proposed Philadelphia customers to request early gas service. In mid-January 1948, the Philadelphia Gas Works, feeling a severe demand on its existing manufactured-gas facilities, requested emergency gas deliveries to begin on February 9.

In response, Texas Eastern attorney J. Ross Gamble sent Charles Francis a copy of the Gas Works' application and noted that "it may well be that the Power Commission will dispose of it [the application] at an early date. It would therefore seem desirable for Texas Eastern to make up its mind as to the position it desires to take, at the earliest possible moment."[52] The company did not have excess gas deliverability and was not even meeting contracted demand.

Two weeks later, Gamble telegraphed David T. Searles of Vinson, Elkins, Weems & Francis that the FPC, late the previous night, had set a hearing on the Gas Works application for February 9. The notice of hearing indicated that after the Gas Works had filed its request for emergency service, other current customers had intervened in the case as well. The proposed hearing set off a mad scramble among customers and other companies interested in gas service. Fearing that the FPC might divert its own gas supplied by Texas Eastern, the Public Utilities Commission of Ohio, the Ohio Fuel Gas Company, the Manufacturers Light and Heat Company, United Natural Gas Company, and Equitable Gas Company, among others, filed informal protests with the FPC to prevent Texas Eastern from diverting any existing gas sales volume to the Gas Works. However, seeing an opportunity to get gas itself, the New York Public Service Commission and the Iroquois Gas Corporation, gas companies doing business in western New York, petitioned to intervene and requested the excess gas.

R. H. Hargrove immediately telegraphed back to Gamble that "we will require a minimum of two weeks to take necessary steps to be in position to deliver gas to Philadelphia Gas Works in the event we are ordered so to do. Believe it would be well for you to clear this up in informal discussion."[53] Gamble confirmed to Hargrove that he had notified the chairman of the commission, the supervising commissioner, and commission counsel of Texas Eastern's position. But an attorney for the Gas Works discussed his client's gas needs with Gamble on February 6 and stated that they were now requesting 4,300 mcf/d for the balance of the winter, presumably because 700 mcf/d of Texas Eastern's total 5,000 mcf/d dedicated for emergencies had already been granted to the Waynesburg Home Gas Company. On February 20, the FPC denied the Philadelphia Gas Works application. The agency seemed content in this case to wait for the emergency shortage to disappear before shuffling gas to other needy customers.[54] Confusion continued over exactly when Texas Eastern would be prepared to deliver gas on a regular basis to both Philadelphia utilities. On June 1, 1948, the Philadelphia Gas Works filed an additional petition with the FPC for a declaratory order to remove uncertainty over Texas Eastern's ability to deliver gas to it under terms of the original contract.[55]

Texas Eastern responded ten days later. The company supported its original proposed service date to the utility by quoting from one of the utility's own letters in which Hudson Reed, president of the Gas Works, wrote that the utility "is entitled to receive natural gas from Texas Eastern Transmission Corporation on July 1, 1948 or as soon thereafter as Texas

Eastern's deliverability reaches 340,000,000 cubic feet."[56] Texas Eastern also noted that it should not have to cease deliveries of gas to its current customers in order to supply gas to the Gas Works.

Finally, on September 17, 1948, Texas Eastern began delivering natural gas to Philadelphia Electric. The utility received its first supply of gas at its Tilghman Street gas plant and used it for reforming and enriching its current manufactured-gas production. After Texas Eastern began supplying natural gas to the utility, Philadelphia Electric's president, Horace P. Liversidge, said, "In view of the large number of customers applying for gas heating services, the supply of natural gas to Philadelphia Electric will permit it to increase its capacity and bring nearer the time when all applicants for gas services may be accepted and adequately supplied."[57] The company actively pursued that goal.

In 1951, Philadelphia Electric began converting its gas system to natural gas. After conducting many studies, the company decided to convert its gas distribution system to mixed gas, a combination of low-Btu manufactured gas and high-Btu natural gas. The company made this decision for several reasons. First, it could not be certain that any contracted supply of natural gas would be always available, but its manufactured-gas plants could be producing gas continually. Second, the manufactured-gas plants could be kept in the utility's rate base only if they were actually producing manufactured gas. However, the switch to mixed gas indicated that a conversion to straight natural gas would come in the future. To produce mixed gas, the gas works raised the heating content of manufactured gas, typically about 530 Btu, to 820 Btu with the addition of natural gas, which has an average heating content of about 1,020 Btu.

Philadelphia Electric did not have to make adjustments in its transmission or distribution system to accept mixed gas, but it did have to adjust all its customers' gas-burning appliances to accept the higher burning Btu gas. The company hired a field force of 400 men to go door-to-door to make the necessary adjustments to each of the utility's residential consumers' gas appliances so that the appliances would accept the higher-Btu gas. Some 170,000 customers required 380,000 adjustments for a wide variety of appliances. The field workers adjusted 220 different types of gas ranges, 160 water heaters, and 100 house heaters, not including commercial or industrial gas appliances. The increasing demand for natural gas led Philadelphia Electric and the city's other utilities to contract for more gas from Texas Eastern and other pipeline companies.[58] This conversion process foreshadowed much larger ones to come in the Northeast.

Table 5.2. Gas Sales of Utilities in Pennsylvania, 1941–1959
(millions of therms)

Year	Natural Gas	Manufactured Gas
1941	1,332	183
1943	1,615	223
1945	1,485	240
1947	1,752	290
1949	1,876	326
1951	2,619	125
1953	2,753	24
1955	3,241	16
1957	3,673	8
1959	4,110	2

NOTE: A therm is equivalent to 100,000 Btu.
SOURCE: American Gas Association, *Historical Statistics of the Gas Industry* (Arlington, VA: AGA, 1964).

By the end of the decade, the manufactured-gas business in Philadelphia had virtually disappeared. Although Texas Eastern was the first major pipeline to sell Texas and Louisiana gas into Philadelphia, others constructed later delivered natural gas to the local utilities by the mid-1950s. Still, the 57 mmcf/d received by the company was not sufficient to eliminate manufactured gas from the company's operations. The company would not convert its system to straight natural gas until the end of the decade. By the mid-1950s, it was clear that manufactured-gas consumption in Pennsylvania was on a permanent decline (see table 5.2). However, these statistics also show that natural gas was rapidly finding new markets; the increase in natural gas sales far outpaced the decrease in manufactured-gas sales.

The Big and Little Big Inch Lines provided Texas Eastern with a path to Philadelphia, which became the first major northeastern city east of Appalachia to convert to southwestern natural gas supply. The Inch Lines represented a ready-made pipeline system capable of transporting southwestern gas to Philadelphia and beyond. An assortment of competitors, including natural gas, oil, coal, and railroad interests, entered the fray. Coal and railroad companies most vociferously opposed the natural gas Inch Lines, but they could not prevent them from delivering a more efficient fuel to beckoning markets.

Political entrepreneurship facilitated this economic process. The primary question revolved around the efficacy of introducing natural gas into the coal-rich and coal-dependent northeastern states. Both sides of the debate hired lobbyists, some of whom occupied high positions in government. The lobbyists' function was to help determine which fuel would prevail and which group of entrepreneurs would profit from industry expansion. Lobbying occurred in federal, state, and local agencies, all of which had to approve gas transmission before gas could flow. Ultimately, the lobbying function was secondary. The key to the process of change was natural gas's clear superiority over manufactured gas and coal. Once natural gas was available at a competitive price, its victory in the marketplace was assured. Natural gas's first assault on manufactured gas paved the way for other pipeline ventures to reach more deeply into the Northeast.

6. The Conversion of New York City

THE NEW YORK City metropolitan area was the largest manufactured-gas market in the United States. Located approximately 100 miles from Philadelphia, it was the next major gas market northeast of Philadelphia. Although pipeline entrepreneurs had expressed interest since the early 1940s in selling natural gas to the New York area utility companies, they found local utility companies wary. During the war, both the FPC and WPB believed that Appalachian-area customers should receive necessary gas supplies before certifying expansion projects farther into the northeast. But after the war, Texas Eastern's success in attaining FPC certification to sell gas to two Philadelphia distribution companies indicated that regulatory authorities would not prohibit gas sales even deeper into the Northeast. As the cost of manufactured gas increased, the regional utilities considered more seriously the possibility of contracting for natural gas.

NEW YORK CITY AND GAS CONSUMPTION

The New York City area had an extensive and well-entrenched manufactured-gas industry. The local manufactured-gas industry developed very early in the New York metropolitan area, and New York City alone consumed about 40 percent of all manufactured gas used in the nation. Residential gas consumers burned most of this gas. For natural gas to break into this huge and lucrative market, a pipeline company had to convince federal, state, and local regulatory agencies, as well as the utility companies themselves, that it could arrange twenty-year contracts for large quantities of natural gas at prices competitive with manufactured gas.

All the utility companies operating in the New York City area distributed manufactured gas. They included some of the nation's largest utility companies, such as Consolidated Edison, Brooklyn Union, Public Service of New Jersey, and Long Island Lighting Company. At least one of these utilities, Consolidated Edison, had considered purchasing natural gas for its distribution system as early as the 1920s. Late in that decade, Consolidated Edison hired Ralph E. Davis, a consulting geologist, "to study the feasibility of a pipeline that might be built from that area [eastern Kentucky and West Virginia] to take gas into New York." Davis's study concluded that a fifteen-year supply of natural gas existed in the Chattanooga shale in eastern Kentucky and West Virginia suitable for use in Consolidated Edison's system. But the Great Depression halted all Con Edison's plans to build such a line. In the early 1930s, after the discovery of the Oriskany Trend in northwestern Pennsylvania and southwestern New York, Consolidated Edison hired Davis again to study the feasibility of building a pipeline from these fields to New York City. The Oriskany discoveries flooded the Appalachian gas market, and the price of gas subsequently fell from a maximum field price of $0.35 per mcf to a low of $0.06 per mcf. But Con Edison's managers feared that the Oriskany reserve life was limited and therefore not dependable for a long-term supply of gas. According to Davis, "This gas supply did not promise the longevity needed to justify Consolidated Gas Company changing over to natural gas."[1]

During the late 1930s and early 1940s, Hope Natural Gas and others recognized the potential for profit in selling natural gas to the New York City area. They formed the Reserve Gas Pipeline Corporation to sell excess Texas Gulf Coast gas to northeastern customers, particularly those in New York. Ray Fish, a designer of the proposed line, recalled that the New York market "was all manufactured [gas] and we could see that it was a kind of ailing industry. . . . We tried to sell that deal [Reserve Gas Pipeline] to the New York market . . . but we didn't make any progress."[2] The Reserve Gas group contacted Consolidated Edison, Brooklyn Union, as well as utilities in Philadelphia offering them delivered natural gas for $0.23–0.25 per mcf, a price competitive with manufactured gas (see table 6.1). The utilities countered with a maximum purchase price of $0.18 per mcf; this would not support the promotion of the pipeline venture. After the United States entered World War II, the Reserve Gas plan was suspended and then dropped.

Table 6.1. Manufactured-Gas Production Costs of One Northeastern Company
(cost per mcf of 550-Btu gas produced)

Production Component	1939	1943	1946	1948
Gas generator solid fuel	$0.014	$0.039	$0.068	$0.063
Oil for generator gas	0.026	0.076	0.120	0.203
Coke oven and other gas	0.087	0.089	0.064	0.144
Other production items (ie., labor, maint., etc.)	0.038	0.051	0.065	0.078
Total	$0.165	$0.255	$0.317	$0.488

SOURCE: American Gas Association, *Gas Rate Fundamentals* (New York: AGA, 1978), 9.

Interest in the New York City gas market revived again after World War II, especially during the debate over the postwar use of the Inch Lines. At the hearings before the Special Committee Investigating Petroleum Resources in November 1945, several witnesses discussed the potential for using the Inch Lines to transport gas into northeastern areas, including New York City. One of these speakers, future Texas Eastern organizer E. Holley Poe, stated that the New York, northern New Jersey, and Philadelphia areas needed approximately 65 bcf per year of natural gas to replace petroleum, but not coal, in the manufactured-gas production process. After winning the bid for the Inch Lines, Texas Eastern's original interest in the New York market continued. But the company determined after taking customers in Appalachia and Philadelphia that it was not immediately prepared to contract for large gas sales into the New York market.

THE FORMATION OF TRANSCONTINENTAL GAS

Other entrepreneurs also had focused their interest on the New York market. Claude Williams, bespectacled and nearly bald at his still young forty-two years, hoped to sell gas into the Northeast through either the Inch Lines or a new system. A former assistant secretary of state of Texas, attorney, and self-described independent oil and gas operator, Williams perceived the Inch Lines as a vehicle to finally make it big.[3] On February 16, 1946, Williams and his uncle, Rogers Lacy, jointly formed Transcontinental Gas Pipe Line Company, Incorporated, of which Williams was the president, to acquire the Inch Lines. Soon after the company was formed, two other associates, Alfred C. Glassell and Alfred C. Glassell,

Jr., joined Williams and Lacy in order to market their own gas reserves to northeastern customers. Later, Tennessee Gas founders Ray Fish and Clyde Alexander left Tennessee Gas and joined Williams's new venture. Fish, the debonair designer of Tennessee Gas and the unsuccessful Reserve Gas, brought his ideas for a similar line to Williams to be used for the construction of a new one if their effort to acquire the Inch Lines proved unsuccessful.[4]

Williams, however, planned first an attempt to purchase the Inch Lines. He formally declared his interest in the Inch Lines, or an alternative new line, on March 1, 1946, when he applied for a certificate to construct and operate a pipeline from the Southwest to New York.[5] Williams also offered to purchase the Inch Lines, and the Southwest Emergency Pipe Lines, a much smaller war surplus pipeline system, from the War Assets Corporation, predecessor agency of the WAA, and convert them to gas transmission. Neither the WAC nor the WAA acted on Williams's informal offer.

In the same application, Transcontinental stated that if unsuccessful in purchasing the government pipelines, it would construct a 26-inch line from Corpus Christi, Texas, to Pennsylvania, New Jersey, and New York. This new line would have an initial capacity of 300 mmcf/d, with the possibility of looping the line with an additional 26-inch system as the market developed. Williams's interest in the Inch Lines continued throughout 1946, and he formally bid on them during the first round of bidding in July. After these bids were thrown out by General Littlejohn, Williams continued his efforts to purchase the Inch Lines, and he also planned contingently to build a new pipeline system in case his second bid for the Inch Lines proved unsatisfactory.[6] Williams's group recognized quickly that the New York area comprised a high demand for natural gas, and either the Inch Lines, or a new system, could serve that market. His strategic decision to simultaneously bid for the Inch Lines and promote a new system marked his determination to enter the industry with a major pipeline system. Within the company, he was perceived primarily as a lobbyist, but he also had an entrepreneurial vision.[7]

On December 11, 1946, Transcontinental filed its first amended certificate application based upon its original application of March 1, 1946. In the amended application, the company provided a more detailed description of a new line and estimated that it could construct its own 1,380-mile, 26-inch system for $130 million based on a proposed gas sales price of $2.25 per therm.[8] It would have a capacity of 325 mmcf/d, about 25 percent of which would consist of flare, or residue, gas. The line would extend from Texas through Arkansas, Missouri, Illinois, Indiana, Ohio, and Pennsylvania to a point near the Hudson River, and the company reported that it had already negotiated for gas contracts with eight utilities in New York, Pennsylvania, Maryland, Delaware, and New Jersey.[9]

The certificate application also addressed the continually troublesome issue of manufactured-gas displacement:

> The natural gas supplied to these companies will be used mainly for enrichment and reforming in connection with their manufactured gas product. The area is presently being served with various combinations of carbonated water gas, coke oven gas and refinery gas. Applicant has agreed with the utilities serving the territory to expand the market for domestic and superior use of natural gas. In negotiations with utility companies Applicant and the utility companies have tentatively agreed upon a contract whereby the only gas not used for enriching or reforming purposes in connection with their manufacturing and distributing business will be used by them during off heating season in their electric generating stations.[10]

Thus, Transcontinental carefully indicated that its gas supply would not be used by utilities to displace their manufactured-gas sales, particularly during the winter heating season.

Less than two weeks after the company submitted its amended application to the FPC, the New York *Journal of Commerce* published an article confidently predicting that New York would soon receive gas supplies. The writer noted that as much as 1 bcf/d of natural gas was still being wasted in Texas due mostly to careless oil production techniques. Abundant southwestern gas supplies and increasingly interested northeastern utilities assured investors that natural gas pipelines were attractive investments. Rapidly increasing costs associated with maintaining and expanding manufactured-gas facilities in the New York region clearly supported this view. During the 1946–47 winter, Brooklyn Union halted home heating equipment sales to residential customers because of inadequate supplies of manufactured gas. According to a New York business writer, abundant southwestern natural gas would alleviate "the need for plant expansion at present inflated costs of construction" and reduce "dependence on bituminous coal and the United Mine Workers."[11] The rising costs of manufactured-gas production, combined with excess natural gas availability in the Southwest and an existing pipeline system connecting the two regions, promised a future of natural gas for New York City.

OPPOSITION TO NATURAL GAS IN NEW YORK

Not unexpectedly, Transcontinental's amended certificate application evoked intense opposition from the solid-fuels industry, railroads, and other natural gas pipeline interests. Eastern Gas and Fuel Associates filed a petition to intervene on December 27, 1946. The firm's attorney, John Gage, cited his client's interest in the market areas of its wholly owned subsidiaries, the Philadelphia Coke Company, the Connecticut Coke Company, the Boston Consolidated Gas Company, and the Old Colony Gas Company, all of which either produced or sold manufactured gas in Pennsylvania and much of New England. The Eastern States Retail Solid Fuel Conference petitioned to intervene. The conference represented the Pennsylvania Retail Coal Merchants Association, Baltimore-Maryland Coal Exchange, New England Fuel Dealers Association, Fuel Merchants Association of New Jersey, Retail Fuel Institute (of Boston, Mass.), New York State Retail Solid Fuel Merchants Association, Delaware State Coal Club, the Coal Dealers Association of Philadelphia, Office of the Coordinator of the Retail Solid Fuel Industry of the City of New York, and a total of 4,900 retail solid-fuel dealer members of these associations. Through its attorney, the conference stated that natural gas importation would result in "serious injury to the anthracite industry and to many thousands of individuals dependent upon it."[12] The Anthracite Institute filed a similar petition with a straightforward message: natural gas should not be introduced into areas which could be adequately served by the solid-fuel industries.[13]

Along with the coal companies, railroads joined the anti–natural gas fracas. Alfred S. Knowlton, representing twenty-three railroads including the Baltimore and Ohio Railroad Company and the Pennsylvania Railroad Company, objected to a Transcontinental certificate on the grounds that natural gas would displace "substantial quantities of anthracite and bituminous coal and coke and other solid fuels upon which petitioners largely depend for traffic and revenue." He also stated that the substitution of natural gas for anthracite and bituminous coal would be "unnecessary and uneconomical. . . . [Natural gas] . . . will substantially reduce petitioners' revenues and to that extent impair their abilities to continue to function efficiently and economically as common carriers in the public interest as required by law."[14] Other opposition to Transcontinental's application included bidders for the Inch Lines, the Organization Committee, Mutual Cooperative Plan of the American Public Utilities Bureau and Big Inch Natural Gas Transmission Company, as well as an existing firm, Southern Natural Gas Company.[15]

Many other petitions to intervene came into the commission as well. The Pennsylvania Public Utilities Commission gave notice of its intention to intervene. Frank Harper, executive secretary of the Public Service Commission of Maryland, requested that the FPC keep his office informed of any developments regarding Transcontinental's proposed service to Maryland. All these intervenors were focusing their attention on the pipeline's application even before the WAA had sold the Inch Lines to the highest bidder. But the Inch Lines, an existing system extending into New York, posed a much more immediate threat to the northeastern coal, oil, and railroad interests. And as soon as Texas Eastern won the bid, the various intervenors in Transcontinental's application refocused their opposition on the new Texas Eastern system.

CREATING A REGULATED FIRM

Having lost the bid for the Inch Lines, Claude Williams redoubled his efforts to construct his own pipeline system. Some of his original partners, however, decided to sell back their stock in the venture. Interest among other investors remained after Ray Fish, who had formed the Fish Engineering Corporation in 1946, became a substantial stockholder and proposed to construct the new line. Although not the powerhouse represented by Texas Eastern founders George and Herman Brown, Fish brought much experience in gas pipeline engineering and design to the proposed construction of Transcontinental's line.

More than anything else, Transcontinental needed an FPC certificate as fast as possible in order to remain competitive with Tennessee Gas and Texas Eastern. On February 17, 1947, Williams requested from the FPC "an immediate hearing on its application to construct a gas transmission line from Texas to the eastern seaboard. . . . Please advise date of hearing as soon as possible."[16] Leon Fuquay, secretary of the FPC, pointed out to Williams that before hearings could take place, Williams's company needed to file "all applicable exhibits listed in part 57.6 (order No. 99) of the commission's regulations under the Natural Gas Act."[17] After receiving and reviewing those documents, the FPC would advise Williams of any dates set for a hearing. Williams, anxious to move ahead, believed that he could sell gas to New York customers for $0.06 per mcf less than Texas Eastern, but first he had to comply with FPC certificate requirements and build a pipeline system.[18]

At the same time, New York regulatory officials were becoming increasingly concerned about gas shortages within their state. Some of the less populated and industrialized regions of the state did consume locally produced natural gas from small local distributors. One of these companies, the New York State Natural Gas Corporation, planned to sell 15 mmcf/d, and 30 mmcf/d later, to a Canadian company. The New York Public Service Commission (NYPSC), the state's utility regulatory agency, opposed the plan because of New York's developing gas shortage problems, in which at least one utility curtailed customers due to a shortage of supply. These and related developments encouraged the NYPSC to support the sale of southwestern-produced natural gas into New York to augment the state's existing albeit limited production and supply.[19]

Transcontinental's intense efforts to build a new pipeline were mired in uncertainty. Although on the surface it seemed that a new gas pipeline was needed and feasible, Claude Williams confronted difficult obstacles. He described his company's position: "Owners of gas were slow to commit reserves to a new company. Distributors of gas wondered about signing up with a company whose line was still on paper. Investors had to be persuaded that the whole gas industry was as good as it looked and that this particular project would pay out. Large diameter steel pipe and plate from which to make it were really scarce."[20] Transcontinental eventually signed numerous gas supply contracts for takes from fifty-four separate gas fields, many of which were located on the Louisiana coast. The company contracted for steel from the Kaiser Company, Incorporated, and Consolidated Western Steel Corporation, and the pipeline essentially paid for the steel with notes and stock.[21]

In addition, Transcontinental began preparing for the FPC certificate hearings. The company amassed a multitude of data to prove to the FPC that it could successfully finance and build its line and supply its large metropolitan distribution customers with an adequate supply of gas. However, there was no guarantee that the FPC would grant a certificate, and potential competitors were also preparing to block Transcontinental's certificate application.

Texas Eastern, for one, was adamantly opposed to Transcontinental's certificate application, and it asked the FPC to delay the application for as long as possible, if not simply to dismiss it. Having successfully bid on the Inch Lines, Texas Eastern feared that another pipeline might capture New York City, the most prized northeastern gas market. During May,

Texas Eastern petitioned the FPC to intervene in Transcontinental's certificate application. In a brief filed by Texas Eastern attorney J. Ross Gamble, the company stated that Transcontinental's service "will largely duplicate the service which can be rendered from the existing facilities of the Intervenor . . . and would constitute a hazard to the successful operation of the Intervenor's system to the detriment of the Government of the United States."[22] Thus, Texas Eastern's posture reflected the idea that if Transcontinental's future operation adversely affected Texas Eastern's profitability and the company suffered financial difficulties, the government might in some way have to bear Texas Eastern's resulting burden.

Time was important in the race for the New York market. And both Tennessee Gas and Texas Eastern were eager to beat Transcontinental to New York. Tennessee Gas, though, intended to serve Buffalo, New York, instead of the New York City area. Tennessee Gas's inability to obtain the necessary twenty-year gas supply contracts for the New York City market prevented negotiations with those utilities.[23] But Tennessee did receive FPC approval to extend its pipeline to Buffalo, New York, and eventually supplied several upstate New York utility customers. Texas Eastern, with its system's proximity to New York City, discussed gas sales with utilities in New York as well as in other northeastern areas. Soon after winning the bid, E. Holley Poe and Reginald Hargrove, who in 1947 became president of the American Gas Association, began meeting with potential customers in Philadelphia and New York.

Thus, while opposing Transcontinental in the regulatory arena, Texas Eastern attempted to contract with New York area utilities, in part to limit its competitor's potential range of customers. In late May, nearly one month after Texas Eastern began operating its pipeline, W. E. Bolte, assistant vice-president of Brooklyn Union Gas Company, wrote a letter to Poe that was subsequently referred to Hargrove. The letter expressed Brooklyn Union's interest in purchasing natural gas from Texas Eastern. Later, in New York, the two Texas Easterners met with Bolte. Hargrove expressed his interest in selling gas to Brooklyn Union. He emphasized, however, that Brooklyn Union, not Texas Eastern, needed to investigate all legal, technical, and engineering problems regarding how Texas Eastern would extend its line to Brooklyn Union. In particular, Hargrove requested information on the feasibility of extending Texas Eastern's pipeline across the Narrows—the preferable and more economical route—or crossing the Hudson River north of New York City and then bringing the pipeline through the city. "In either event . . . ," Hargrove wrote, "the question of the availability of pipe will be of paramount importance."[24] In fact, Hargrove's insistence that Brooklyn Union conduct the investigations regarding river crossings for the pipeline and the availability of steel may well have proved to be a tactical mistake. The provider, Texas Eastern in this case, was asking the customer to do all the necessary research for a profitable service Texas Eastern would perform; for its part, Brooklyn Union kept its gas supply options open.

During Texas Eastern's FPC certificate hearings in the summer of 1947, Malcolm F. Orton of the Public Service Commission of New York, testified on his state's interest in natural gas. Orton stated that certain cities in New York experienced shortages of natural gas similar to those experienced by Philadelphia. However, Orton requested that the FPC not certify Texas Eastern to sell gas into Philadelphia until the Appalachian fuel shortage ended. Only then, Orton said, should the FPC permit new gas service in the other northeastern areas. And at that point, he noted, an "opportunity should be afforded to present the claims of other territories than those proposed to be contracted for by the applicant [Texas Eastern], particularly New York City and vicinity."[25] Texas Eastern opposed Orton's contention that it should serve only Appalachian customers until the shortage was over. The FPC, believing that some portion of Texas Eastern's capacity should be dedicated to new markets, did certify Texas Eastern to sell gas to customers in Philadelphia. Once again, however, Texas Eastern's interests were opposed to those of the potential New York area customers. These situations may not have helped the company's long-term strategy of selling gas to New York City. At the same time, Transcontinental engaged in contract negotiations with the New York City utilities.

TRANSCONTINENTAL'S CERTIFICATE HEARINGS

After Transcontinental filed the necessary documents with the FPC, the regulatory agency scheduled public certificate hearings to begin on October 27, 1947. For the hearings, a total of fourteen intervenors supported Transcontinental's certificate application. All of them were either northeastern utilities or public service commissions anxious to have natural gas supply in their areas. Opposing intervenors included the expected panoply of coal, railroad, and labor organizations which feared that the introduction of natural gas into the New York metropolitan area would have serious economic effects on their business. Several gas pipeline companies including Texas Eastern opposed the application as well.

At the hearings, Transcontinental proposed to construct a transmission line consisting of 1,760 miles of 26-inch pipe, including 80 miles of additional line, and fifteen compressor stations with a total of 160,000 hp to transport 340 mmcf/d. It now estimated the total cost of constructing this system to be $151,380,426, or $20 million more than the company bid for the Inch Lines.[26] A Transcontinental representative maintained that the majority of its gas sales would displace petroleum, not coal, in the manufactured-gas reforming and enriching process. The company also intended to sell large volumes of natural gas to be used for underfiring boilers in electrical generating plants on an interruptible basis.

Although there was some question as to the availability of sufficient quantities of steel necessary to build the system, a Special Subcommittee on Petroleum of the Committee of Armed Service of the House of Representatives noted that "the provision of steel to those who plan to construct such pipe lines [natural gas] is of No. 1 priority in importance."[27] The subcommittee encouraged any means through which petroleum consumption could be reduced, including increased natural gas use. Although the war had been over for more than two years, the immediate availability of petroleum in case of future conflict stimulated Congress to support all ways to reduce petroleum consumption in order to conserve the fuel for future emergency situations.

The first round of hearings ended on February 21, 1948. The FPC was clearly dissatisfied with certain aspects of Transcontinental's application, particularly in regard to its lack of planning for northeastern gas storage facilities and inadequate gas supply contracts. The next step in the hearings process, which included filing briefs for the concluding oral arguments, allowed at least one of its challengers to take advantage of the company's poor showing during the hearings.

The newly organized Texas Eastern provided some of the most aggressive opposition to Transcontinental's application. Earlier in the year, Texas Eastern's president, Reginald Hargrove, announced his company's intention to build a separate 26-inch pipeline to add 400–450 mmcf/d to the company's capacity. This line, Hargrove stated, would supply gas to New York and lower New England, the same general market area targeted by Transcontinental. During Transcontinental's hearings, Texas Eastern attempted to block the competing project with a motion filed with the FPC to dismiss its application to build the line. A Texas Eastern attorney argued that the line was not necessary because ultimately his company intended to serve the New York area and the newly proposed project would adversely affect the orderly economic development of Texas Eastern's system "and would have a detrimental effect on the potential natural gas consumers in the area to be served."[28]

The FPC denied Texas Eastern's motion and took the opportunity to make an important point regarding competition in the interstate gas pipeline business. The FPC replied: "It is manifest that Texas Eastern seeks to obtain a monopoly of the natural gas markets in the Middle Atlantic seaboard area. We cannot subscribe to the thought that Texas Eastern is entitled to preempt such markets or that recognition of such prospective monopoly is in the public interest. It is therefore concluded that the position taken by Texas Eastern in this regard is without merit."[29] The commission told Texas Eastern that it "should go out and fight for business." The FPC proved in this instance that it would allow and encourage competition for gas markets. The FPC acknowledged here that it was treating the natural gas pipeline industry as one with oligopolistic powers where several of the highly capital-intensive systems could serve the same customer in a competitive business environment. Although the FPC rebuffed Texas Eastern's challenge, the agency remained displeased with Transcontinental's lack of interest in pursuing storage facilities and its inadequate gas purchase plan.

On March 31, 1948, the FPC announced that Transcontinental "had not demonstrated that it possessed an adequate supply of natural gas with which to support its proposed project."[30] The company testified previously during the hearings that it had investigated gas storage facilities in New York and Pennsylvania, and that the inclusion of a storage project in its plans would increase the overall cost of its gas by $0.045 per mcf. However, the FPC reopened the hearings and allowed Transcontinental to present additional evidence to prove that it had commitments for its maximum daily gas requirements for twenty years, and this time the FPC agreed.

As with both Tennessee Gas and Texas Eastern before it, anxious potential customers and interested state public utility commissions greatly aided Transcontinental's application. In 1948, the New York Public Service Commission, which regulated the sale of natural gas by utilities within the state, recommended that gas utilities in New York not already negotiating with Transcontinental intervene in the hearings and request natural gas for their own systems.[31] The NYPSC stated that manufactured-gas companies throughout New York, as well as throughout the nation, were undergoing a rapid inflationary period: Utilities "are faced with operating costs which have increased in some instances to a point where not only is there no return on investment but substantial deficits are experienced."[32] Unexpectedly, Consolidated of Baltimore also made a request for natural gas. Although it had not previously negotiated for gas purchases from Transcontinental, the Baltimore gas distribution utility indicated during the hearings that it was also seeking sufficient quantities of natural gas with which to convert its system from manufactured gas to natural gas. Natural gas, which cost less and had twice the heating value of manufactured gas, promised to be a more economical and efficient fuel.[33]

In considering Transcontinental's application, the FPC acknowledged that the existing northeastern petroleum shortage could be alleviated if natural gas supplies were sufficient to replace the fuel oil used to produce manufactured gas. The FPC encouraged any measures to conserve petroleum products, and the agency estimated that natural gas would displace 629 million gallons of fuel oil per year. The FPC considered this to be a positive step in the alleviation of the ongoing oil shortage rather than an economic hardship to be imposed on the oil industry.[34]

However, the displacement of solid fuels, particularly coke and coal, presented a more controversial problem. Despite Transcontinental's earlier statements that its customers would use natural gas only to enrich manufactured gas and displace the fuel oil used in the manufactured-gas production process, it became clear during the hearings that the new natural gas supplies would also displace significant amounts of coal and coke. The utilities planning to purchase natural gas showed that they would reduce their coke consumption for the production of manufactured gas by 600,000 tons annually. Moreover, Transcontinental proposed to sell 22.5 bcf of natural gas to utilities for underfiring boilers in their gas plants and electrical generating stations. These sales would displace approximately 860,000 tons of coal during the pipeline's first year of operation.[35]

The FPC, on the other hand, noted that "the economic impact of the relatively small and decreasing displacement of coal resulting from the project upon the interests of the coal and associated intervenors should not be serious."[36] According to the figures presented, the distribution companies, by using natural gas instead of oil in the manufactured-gas enriching process, would save as much as $35 million, which would be reflected in lower fuel bills to a total of 4 million customers, a cost and price reduction the FPC could not ignore.

On May 29, after a somewhat tumultuous series of hearings, the FPC finally issued a certificate to Transcontinental. The entire process lasted nine months, required 200 witnesses, filled 9,000 pages, and cost approximately $500,000.[37] The FPC imposed a litany of conditions on Transcontinental's certificate. The FPC required the company to begin actual pipe laying within one year, and it ordered the pipeline company to have its system operational and delivering gas to its customers within twenty-eight months of the May 29 certification date. The FPC also requested quarterly progress reports on construction activities, a tariff consisting of rates, charges, and classifications, among other features, satisfactory to the FPC. Finally, the FPC ordered the pipeline to continue investigating storage facilities for its own system and requested that it consider gas sales to the Columbia system, and others in Appalachia, for their storage systems and to report back within two years. Although Transcontinental's presentation to the FPC in its hearings was barely satisfactory, northeastern demand for gas encouraged the FPC to grant the certificate and guide the company through its construction and initial organization phase.

SELLING NATURAL GAS TO NEW YORK CITY

After the hearings concluded, Transcontinental quickly filed a petition to enlarge its system, even before it had begun construction. The company proposed to use 1,210 miles of 30-inch pipe in place of its original 26-inch pipe. In addition, the company wanted to modify its compressor station system in order to increase the capacity of the pipeline from the original 325 to 505 mmcf/d. The FPC approved the application and modified the line construction schedule so that construction would begin by May 29, 1949, and end no later than April 1, 1951.[38]

Transcontinental Gas broke ground for its line, hailed as the "world's longest natural gas pipeline," on May 23, 1949, in Laurel, Mississippi. Construction began less than one week before the FPC's deadline, but significant pipe-laying activity did not begin until July. The southern segment of the 1,840-mile pipeline path was divided into six 100-mile segments, and this portion was completed by the end of the year. In December, the company began laying pipe on the New Jersey side of the Hudson River and worked southward to meet the system's main line. During 1950, fourteen pipe-laying crews completed the remainder of the

Transcontinental Gas Pipe Line Corporation. Solid black line indicates
pipeline route.

system. Before the line reached New York, the company began its first gas
deliveries to Danville, Virginia, along the pipeline route. Transcontinental, now a $233 million pipeline, expected to be delivering gas to its New
York customers by early 1951 (see map).[39]

Before Transcontinental completed its pipeline, Texas Eastern captured
one of its prospective customers. The relatively small New York and Richmond Gas Company was dissatisfied with its treatment by Transcontinental.
The company originally had to intervene, or appeal, to the FPC for inclusion
among the retail distributors to receive Transcontinental gas and, although
successful, had been granted only 2.5 mmcf/d out of an initial capacity of
340 mmcf/d. Companies such as Consolidated Edison of New York, Public
Service of New Jersey, and Brooklyn Union Gas, although having many
more customers, still received disportionately higher allocations. Further,
those utilities planned to use the natural gas as a substitute for oil in the gas-
enriching process; Richmond Gas wanted to convert entirely to natural gas
use. Its management felt that complete conversion to natural gas would bring
substantial profits to the company and lower costs to the customer. Moreover, since Texas Eastern had transmission lines to the south in nearby New
Jersey, distribution lines to that area, including Staten Island, supplied by
Texas Eastern gas, could be constructed with modest capital outlays.[40]

Consequently, New York and Richmond applied to the FPC to be served by Texas Eastern instead of Transcontinental Gas. The FPC agreed, allowing New York and Richmond to cancel its allocation from Transcontinental. In late August, Texas Eastern became the first pipeline company to deliver natural gas to New York City when it commenced gas deliveries to the New York and Richmond Gas Company, which served Staten Island, the Richmond borough of New York City. In a ceremony marking the historic event, New York's mayor, William O'Dwyer, lit the first natural gas flame on Staten Island produced from gas supplied by Texas Eastern. In the first full year of operation with natural gas, New York and Richmond earned a net profit of $200,000 after a loss of $45,000 in 1948. Moreover, during 1949 the company passed on a 11 percent rate reduction to customers. Although Texas Eastern captured a New York City area utility before Transcontinental began delivering gas into the area, its prize was more symbolic than anything else. Nonetheless, Texas Eastern served notice that its initial exclusion from the larger area utilities did not necessarily mean a cease-fire in the ongoing competition for new northeastern customers.

Once certified, constructed, and operational, Transcontinental quickly became a competitive factor and one of the top three pipelines vying for a share of the northeastern market (see table 6.2). It was not only preparing to deliver gas into the New York City area, it had begun selling gas to the Philadelphia Electric Company, one of Texas Eastern's prized customers, and exemplifying in the process the FPC's policy of treating the natural gas industry as a natural oligopoly. Although Transcontinental's gas sales of 57 mmcf/d to Philadelphia Electric "was nearly three times the amount received in 1950 [from Texas Eastern], it was not nearly enough to warrant a complete changeover to natural gas."[41] In fact, most of Transcontinental's gas was reserved for its New York customers.

Despite the heated competition between Texas Eastern and Transcontinental for New York, cooperation among the growing network of gas pipelines was essential for the successful operation of the gas industry as a whole. Late in December 1950, as Transcontinental was preparing its line for operation, a work stoppage at Public Service of New Jersey's manufactured-gas plants created a system-wide gas shortage. Transcontinental then contracted for emergency deliveries from Texas Eastern to purge and pressurize the trunk line so that it could begin delivering natural gas early to Public Service of New Jersey. Later, Texas Eastern agreed to sell part of

Table 6.2. Transcontinental Gas Pipe Line Corporation Operations, 1951–1954

Year	Gas Sold and Transported (bcf)	Operating Revenues ($ million)	Miles of Pipeline	Personnel	Reserves (tcf)
1951	136	39	1,200	1,419	4
1952	183	52	1,284	1,284	4
1953	192	59	1,832	1,241	4
1954	198	63	2,482[1]	1,287	6

[1]Includes some proposed mileage.

SOURCES: Transcontinental Gas Pipe Line Corporation, *Annual Report* (various years), and Moody's Public Utility Manual.

its Oakford natural gas storage facility to Transcontinental so that it could comply with the FPC's insistence that it acquire such facilities.

CONVERTING NEW YORK CITY TO NATURAL GAS

Transcontinental's sole grip on Consolidated Edison, Brooklyn Union Gas Company, and other area utilities remained firm at least for the short term (see table 6.3). Gas demand was concentrated in five major New York gas companies, Con Edison, Brooklyn Borough Gas, Brooklyn Union Gas, Kings County Lighting Company, and Long Island Lighting Company, which all contracted for purchases of natural gas from Transcontinental under contracts dated July 25, 1950, superseding contracts signed by all the same parties during 1948.

Under the terms of the New York facilities' agreement, these utilities jointly agreed to take responsibility for natural gas deliveries from Transcontinental's terminus at the east bank of the Hudson River at 134th Street. The five utilities agreed to cooperate in financing and constructing several underwater pipelines from Transcontinental's terminus to points in New York City where the utilities would receive their gas. They also agreed to construct separate lines from the receiving point to connect with their own distribution systems. Consolidated Edison was the largest purchaser in this group of five utilities, and it accounted for approximately 20 percent of Transcontinental's total gas sales volume.[42]

The five utilities agreed to construct those sections of the line that ran through their own franchise territory. Some portions of the line were used by two or more of the five companies; in those cases all costs were apportioned appropriately. The utilities estimated the total cost of the

Table 6.3. Transcontinental Gas Pipe Line Corporation's Gas Sales

Customer	Contract Volume (mcf)
Consolidated Edison Company	128,000
Public Service Electric Gas Company	70,000
Brooklyn Union Gas Company	70,000
Northeastern Gas Transmission Company	64,000
Philadelphia Electric Company	37,000
Long Island Lighting System	35,000
Philadelphia Gas Works	25,000
Piedmont Natural Gas Company	20,000
South Jersey Gas Company	20,000
Elizabethtown Consolidated Gas Company	15,000
Others	69,657
Total	553,657

NOTES: These are Transcontinental's original proposed contract sales amounts. Actual amounts changed after the line began operations.
SOURCE: Transcontinental Gas Pipe Line Corporation, *Annual Report* (1950), 20.

262,830-foot line to reach $14 million. It contained sections of pipe measuring from 12 to 30 inches in diameter with a working pressure capacity of 350 psi. Because the high-pressure line would travel under heavily traveled metropolitan streets in densely populated areas, the pipeline had high safety standards. Owing to the line's high cost and expensive safety precautions, gas industry observers dubbed it the "Safest Inch" as well as the "Costliest Inch."[43]

Transcontinental's two major New York City customers, Consolidated Edison and Brooklyn Union, began receiving natural gas on January 16, 1951. The two utilities accounted for gas deliveries to nearly all the residential gas consumers in the New York City area, and their eventual conversion from manufactured gas to natural gas represented a substantial loss to the manufactured-gas industry as a whole (see table 6.4).

Consolidated Edison, a gas and electric utility, served more than a million gas customers, mostly residential. Brooklyn Union served approximately 350,000 gas customers in the boroughs of Brooklyn and Queens. Despite Texas Eastern's apparent coup in selling the first natural gas to New York City when it began supplying gas to the New York and Richmond, the New York media recognized Transcontinental as the first substantial supplier of gas to the City. The *New York Times* noted that "while this is not the first natural gas to get here — Staten Island has had it since 1949 — it is the first to arrive in any volume."[44]

Table 6.4. Conversion from Manufactured Gas to Natural Gas:
Two New York Utilities
(bcf)

| Year | Brooklyn Union Gas Company[1] | | Consolidated Edison Company of New York[2] | |
	Natural Gas Produced	Manufactured Purchased	Natural Gas Produced	Manufactured Purchased
1948	35[3]	—	58	—
1949	33[3]	—	57	—
1950	36[3]	—	62	—
1951	38[3]	14	58	3
1952	43[3]	22	50	29
1953	—	25	29	52
1954	—	31	39	57
1955	—	31	15	57
1956	—	38	4	60
1957	—	47	—	69
1958	—	55	—	82

NOTES:

[1]Brooklyn Union began its conversion during early 1952 and completed the conversion on August 27, 1952.

[2]Consolidated Edison began its conversion during April 1950 and completed the conversion during the summer of 1956.

[3]These volumes are sales volumes.

SOURCE: Moody's Public Utility Manual, various years.

Consolidated Edison's long-range plan called for converting its gas distribution system entirely from manufactured gas to natural gas. But this plan was not based on years of exhaustive research. Instead, "changing economic conditions," particularly the rapidly increasing cost of manufactured gas and the availability of ample supplies of natural gas, led the company to the obvious choice. The decision to convert to natural gas was not without short-term costs. Con Edison had a $250 million investment in gas-manufacturing plants, plants which were becoming obsolete and expensive to operate. Natural gas offered a cost-competitive fuel that was much more efficient and clean-burning than manufactured gas. After World War II, according to a management study of Consolidated Edison's decision-making process, "natural gas [had] a clear and unmistakable cost advantage over the traditional manufactured product."[45]

Consolidated Edison's growing interest in natural gas was a harbinger of the nation's largest metropolitan fuel conversion. The entire regional

Table 6.5. Gas Sales of Utilities in New York and New Jersey, 1945–1959
(millions of therms)

Year	Natural Gas	Manufactured Gas
1945	162	821
1947	222	963
1949	343	995
1951	670	836
1953	1,307	364
1955	1,822	251
1957	2,589	96
1959	3,491	58

NOTE: A therm is equivalent to 100,000 Btu.
SOURCE: American Gas Association, *Historical Statistics of the Gas Industry* (Arlington, VA: AGA, 1964).

manufactured-gas industry was teetering on the edge of industrial extinction, and a decision by New York's large utilities to convert to natural gas would effectively end manufactured gas's 130-year existence in American energy history. It was ironic that an old synthetic fuel was on the verge of being replaced by a natural one during the age of industrial innovation (see table 6.5).

Many technical problems, however, would have to be addressed before a full-scale conversion could take place. Transcontinental initiated the conversion process by using natural gas to enrich its production of manufactured water gas, which normally had a lower heating content than manufactured coal gas but was less expensive to produce. By mixing it with natural gas, the heating value matched that of manufactured coal gas. This method saved Consolidated Edison money while beginning the overall process of converting the total system to natural gas. It also allowed the utility to keep its manufactured-gas plants in operation while it evaluated the opportunity to switch its system entirely over to natural gas.

Consolidated Edison was assured of a long-term and increasing supply of natural gas. The utility directed its System Engineering Department to prepare studies to guide the proposed conversion of the system to natural gas; the Gas Planning Division was responsible for the actual conversion process. The first priority in planning for the conversion was determining the order in which different parts of Con Edison's service area would be converted to natural gas. After evaluating the layout of gas distribution mains, patterns of demand for gas, and the availability of manufactured gas, the company's Commercial Operations Center and System Engineering Department stipulated that each of the utility's local service areas be converted in the following order: Westchester County, the Riverdale section of the Bronx, the Third Ward of Queens, the First Ward of Queens, the east Bronx, the west Bronx, and Manhattan. These areas comprised approximately 1.4 million gas customers, each of whom operated an average of two gas appliances. Every gas-burning appliance had to be adjusted to burn natural gas since it had a significantly higher heating content than manufactured gas.[46]

Consolidated Edison began the conversion of the first area, Westchester County, in April 1950 and finished it by the summer of 1951. To convert the appliances of the county's approximately 207,000 gas customers, the huge utility hired North American Conversion Company, a firm specializing in natural gas conversions, to undertake the project. North American brought in 600 experienced workers to visit each customer and adjust each gas-burning appliance. Having learned the process of converting appliances from North American, Con Edison itself converted the remaining service areas. By taking over the conversion process, Con Edison was able to employ many of those workers at the manufactured-gas plants that were being mothballed, and the conversion team grew to 1,100 employees during the final stage of its work, the conversion of Manhattan. By the summer of 1956, the utility had converted most of its Manhattan customers to natural gas. Con Edison accomplished the entire conversion of its service area for approximately $36 million. Although several of Con Edison's gas plants were kept in working order for backup, they were rarely used.[47]

Brooklyn Union also underwent a massive conversion program, but only after fully investigating costs of expanding its own manufactured-gas system. During the late 1940s, Brooklyn Union Gas Company instituted a three-year, $25 million construction program to expand its peak capacity for manufactured gas by approximately 10 percent. Yet during these years, the company could not even keep pace with growing demands for gas in its service territory. The northeastern utilities had historically invested hundreds of millions of dollars to build manufactured-gas plants staffed by thousands of factory workers. If this industry was to expand dramatically to meet the postwar demand for gas, its owners would be required to make even greater commitments of funds to build plants and hire new workers. This prospect was not particularly attractive to any gas distributor.

Brooklyn Union first began to consider seriously contracting for natural gas service during the bidding for the Inch Lines in 1946. Brooklyn Union had unsuccessful discussions with Texas Eastern in 1947 about gas purchases, but it was able to contract with Transcontinental for gas. Brooklyn Union officials decided that Transcontinental's line offered it distinct advantages anyway. Whereas Texas Eastern's system traveled through the highly industrialized Appalachian region, Transcontinental's followed a more southerly course, offering Brooklyn Union a gas supply far less likely to be curtailed due to an unexpected, or continuing, Appalachian fuel shortage.[48]

Brooklyn Union knew from studies conducted by Peoples Gas Light & Coke Company of Chicago and the Michigan Consolidated Gas Company of Detroit that utilities which purchased large quantities of natural gas found themselves in a greatly improved operating position. Brooklyn Union first considered using natural gas to produce mixed gas rather than converting directly to natural gas. By utilizing mixed gas, the company could increase its current gas output without jeopardizing its gas supply.[49]

Brooklyn Union's final decision to produce and distribute mixed gas was based on a simple calculation. By producing mixed gas from its proposed 70 mmcf/d allotment combined with its existing 213 mmcf/d production capacity, Brooklyn Union could raise its gas output, after accounting for fuel loss, to 265 mmcf/d with a total investment of about $4 million. A similar increase provided solely by the addition of manufactured-gas production facilities would cost the utility $10 million. The decision to contract for natural gas was an obvious one to make for both short- and long-term economic reasons.[50]

Brooklyn Union undertook a further series of studies of its own system. After a thorough analysis of the economics of using natural gas to produce mixed gas, the company determined that it should not undergo a complete conversion to mixed gas. If the company converted its system to mixed gas and then, as expected, to straight natural gas several years in the future, it would have to adjust its customers' appliances twice and make several costly modifications to its own system. Instead, Brooklyn Union decided to move toward converting its entire system to straight natural gas within the near future.[51]

Brooklyn Union fueled more than 2 million appliances including oven ranges, gas refrigerators, water heaters, various indoor heaters, and other gas appliances, all of which had to be converted from the low-Btu manufactured gas to high-Btu natural gas. The conversion process began on March 6, 1952. The company's residential customers were converted to natural gas within six months. The company's 5,000 industrial customers underwent a similar conversion process to their furnaces. By the end of the year, Brooklyn Union's entire service area was using natural gas.[52]

The conversion of the New York City area from manufactured gas to natural gas was an important event in domestic fuel conversion. Indeed, it marked the end of manufactured gas as a significant industry in the United States. Utilities mounted a tremendous labor-intensive effort to convert their customers' appliances to accept natural gas, and new customers were quickly added to the system. Not only had natural gas displaced an existing fuel, it had, by breaking into the largest gas market in the country, expanded into one of the foremost energy-consuming regions in the United States.

The New York area was the Northeast's prize gas market. Despite losing the bid for the Inch Lines, Transcontinental was the first entrant into this most profitable of all northeastern markets. Transcontinental's promoters were extremely determined to achieve this goal. They were the only low bidders for the Inch Lines not already operating a pipeline system to successfully build a system to serve the large northeastern gas market. Transcontinental's promoters succeeded in acquiring an FPC certificate and persuading the New York Public Service Commission to actively support the introduction of natural gas into the state. Claude Williams's style of political entrepreneurship rested on an impressive promotion of natural gas to New York's manufactured-gas utilities and regulatory commissions.

A combination of federal, state, and local regulatory systems shaped Transcontinental's success. The regulatory process pushed and pulled the entrepreneurial impulse to build the line. Texas Eastern and Tennessee Gas desired to enter the New York market, but the FPC mandated those lines to continue serving their original customers, who more urgently needed natural gas. The New York market represented not only a vast demand for natural gas, it was the last remaining highly concentrated manufactured-gas market. Accounting for nearly 50 percent of all manufactured-gas sales in the nation, the New York City metropolitan area was the prize behind the Pennsylvania coal industry's opposition to northeastern natural gas sales. While Texas Eastern gained access to Pennsylvania markets, Transcontinental, unencumbered with FPC-mandated customers, built a line designed specifically to serve New York City. With access to abundant supplies of the fuel, New York City area utilities were amazingly quick to begin converting their distribution systems to natural gas, effectively ending the century-old domination of the Northeast by manufactured gas.

The success of natural gas in New York did not end competition among the three Southwest to Northeast pipeline companies. The entire New England area remained without access to natural gas, and well before the New York utilities converted to natural gas, an intense competition for New England customers was in progress.

7. New England Spoils

NEW ENGLAND was the last northeastern region lacking natural gas service.[1] One industry observer noted that Boston was the most distant major market to the Northeast, 2,000 miles from the gas fields of south Texas. "Will they ever meet?" he asked. "That is the 64 thousand dollar question."[2] Although the entire region accounted for no more than 50 percent of New York State's manufactured-gas consumption, New England's major cities represented significant and ripe natural gas markets. Besides the economic rewards, capturing New England would be a prestigious prize for whichever pipeline company could win FPC approval to serve it.

An intense regulated competition between Tennessee Gas and Texas Eastern ensued for gas sales into the region; Transcontinental, its "hands full" with New York, remained an interested observer. "It was a lot of fun from where I sat watching," recalled a Transcontinental founder, "these two fellows fighting over this market."[3] Another gas man said at the time, "In all my forty years in the gas business, I have never seen so much chess playing."[4] The battle for New England was fought in both the regulatory arena and the courts. The pipeline company that acquired a certificate would profit from gas sales into the region, but demand for the fuel drove the industry's expansion.

New England was virtually without natural gas in any quantity. Industries used coal and oil to fire their furnaces or generators. Residential customers used manufactured gas or fuel oil. Because of the high concentration of fuel oil consumption in New England, fuel oil dealers were as concerned, if not more so, than coal companies over the introduction of natural gas into New England. In fact, fuel oil and propane use in many smaller and outlying communities was not overtly threatened by natural gas because the expense of constructing a widespread distribution system to small areas would not be profitable. Instead, natural gas companies targeted the larger metropolitan regions, and their respective distribution companies, rather than the entire New England region as a whole.

At various times, Tennessee Gas, Texas Eastern, and Transcontinental planned to expand their pipeline systems to bring natural gas into New England. All three, however, were tied into existing customers in Appalachia, Philadelphia, or the New York City area. Selling gas into New England would require major expansions by each, or all three, of the pipelines. Although Texas Eastern indicated first its interest in the New England market when it bid for the Inch Lines, it did not then have enough gas to serve even its existing Appalachian customers. Tennessee Gas was in a similar situation with one important difference. Working together against Texas Eastern, Transcontinental and Tennessee Gas agreed that Transcontinental would expand its capacity in order to supply gas for Tennessee Gas's New England expansion program. The race to enter the New England market became a competition to acquire a certificate.

Simply contracting southwestern gas supplies for New England customers was not difficult. But contracting for gas sales to New England customers was a different matter altogether. Before New England utilities could feel comfortable, politically and otherwise, receiving natural gas from Texas- and Louisiana-based pipelines, those pipeline companies had to prove themselves capable of reliably supplying natural gas to New Englanders. This task required skills in public relations, industrial strategy, and political entrepreneurship, both in and out of FPC hearing rooms.

"REGGIE VS. MR. PIPELINE"

The competition between Texas Eastern and Tennessee Gas occurred in four overlapping arenas: the public media, the gas industry, the Federal Power Commission, and the lobbyist's den. The public relations competition depended upon the personalities of each company's leader as well as the company's business strategy. The business press as well as the popular press often depicted the ensuing New England competition as a man-to-man duel between Reginald Hargrove of Texas Eastern and Gardiner Symonds of Tennessee.[5]

Contrasts between the two men offered ample journalistic opportunity to exploit their differences. A writer for *Forbes* described the ensuing battle for New England as the match of "Reggie vs. Mr. Pipeline." The fifty-three-year-old Reggie, a large, balding man who was actually referred to as "Reg" by his friends and associates, was a seasoned gas executive. Educated at Rice University in Houston, Hargrove had worked his way up the ranks to second-in-command at United Gas Corporation. Described once as "the finest witness in America," he expressed his belligerence through "tart quips" rather than the "sharp rap of a fist."[6] Symonds, on the other hand, was a Stanford- and Harvard-educated banker from Chicago who typified the newer, openly combative style of gas-pipeline management. He was, one report suggested, "the most aggressive contestant . . . in this industry of scrappers."[7] Still sore at Texas Eastern for having "swiped the Inches right out from [his] compressors," Symonds was eager to settle the score in New England.[8] Contrasts between these two leaders enlivened the battle for New England gas markets, and their ongoing struggles remained a topic of broad general interest in the region and the industry.

The two men first competed against each other in the bid for the Inch Lines. Now Gardiner Symonds had a chance to expand his system into both New York and New England before Texas Eastern. During a trip to New York with one of his attorneys, H. Malcolm Lovett of the Houston law firm Baker & Botts, Symonds expressed concern over his company's ability to contract for sufficient quantities of gas to serve markets in New York. Staying in the Gramercy Park Hotel, Lovett recalled that "Symonds worried about getting [gas] to New York. . . . He was a very emotional sort of man . . . I will never forget. I was sound asleep. There was a tremendous commotion and he bounced out of bed and landed on the floor, and said, 'Malcolm, we're going to New England,' and that's where that started."[9]

Tennessee Gas had been expanding its deliverability capacity even before the Inch Lines had been placed in natural gas service and appeared to be in the best position to capture the New England market. Between November 1946 and October 1947, Tennessee Gas applied three times for FPC approval to increase its capacity to a total of 600 mmcf/d and sell most of the additional gas to existing customers—at a time when the Inch Lines were capable of delivering only 140 mmcf/d. The expansion called for the company to extend its natural gas service into New England by constructing a third main line traveling through Ohio, Pennsylvania, New York, and into Massachusetts, eventually increasing its total system capacity to 1.3 bcf/d.[10]

GARDINER SYMONDS WOOS NEW ENGLAND

With an expansion program in the works, Tennessee Gas's interest in the New England market led its aggressive president, Gardiner Symonds, to embark on a campaign to encourage and prepare New Englanders for natural gas. One of his early and controversial attempts to convince the New England public to support natural gas came early in 1948 during a debate with the ever-present King Coal, Tom J. McGrath. In this debate, McGrath represented the Fuels Research Council, Incorporated, which was dominated by coal and railroad companies. The New England Council sponsored the debate in Providence, Rhode Island. The governors of the six New England states composed the official body of the council. Founded to help reinvigorate the languishing New England economy, the council embodied a certain sense of New England pride and self-sufficiency.

Richard L. Bowditch, president of the council, was also president of a New England coal company. Convening the debate, he stated that despite his interest in coal, "if it can be demonstrated that bringing natural gas into New England will help our economy, I am for it."[11] Bowditch cautioned that New England was located "at the end of the line, fuel-wise" and that the ultimate answer to the natural gas question would have far-reaching effects on New England's economy.

Gardiner Symonds spoke first and presented a broad discussion of the value of natural gas in both industrial heating processes requiring precise heat temperatures and in residential heating and cooking. Symonds acknowledged the opposition to Tennessee's plan to bring natural gas into New England, but he observed that many New England gas distribution companies and industries had expressed interest in the superior fuel. Symonds concluded by arguing that natural gas in New England would displace only a very small percentage of the coal used annually in the United States.[12]

McGrath responded by defending manufactured gas. He noted that the severest gas shortages in Appalachia and New York occurred in areas receiving only natural gas. Manufactured gas could be produced when needed, whereas a community's access to natural gas supply depended upon long-distance pipelines and sometimes unreliable gas wells. McGrath harped again upon the thousands of coal and railroad jobs that he alleged would be lost if pipelines introduced natural gas into New England. Echoing Bowditch's comment that New England was at the end of the fuel line, McGrath stated that during a time of fuel shortages, New England would be the last region to receive the fuel if connected with distant gas reserves. McGrath also attacked the vulnerability of pipelines to normal breakage, and even sabotage, which might leave a community without fuel for hours or days.[13]

Perhaps McGrath's most scathing attack was a personal one against Symonds and all gas industry promoters: "So I say, gentlemen, a promoter (and I hate to use that term 'promoter') a promoter can come along and sell you the supposed advantages of natural gas and put it in here, and he makes a profit on all he sells you—but you better make sure that he is going to be able to sell you all you need and to sell it to you when you need it."[14] Despite McGrath's aggressive attack on natural gas, the superior qualities of the fuel and its increasing availability indicated that it would soon reach New England markets.

E. HOLLEY POE, REGINALD HARGROVE, AND NEW ENGLAND

Texas Eastern was equally aggressive in planning for gas deliveries into New England. Shortly after the company began operations in 1947, E. Holley Poe proposed to organize a New England natural gas distribution company, the Minute Man Gas Company, to supply the New England market. Texas Eastern would become Minute Man's gas supplier and hold the major interest along with three large local distributors. These included Eastern Gas & Fuel Associates, which owned Boston Consolidated Gas Company, the largest New England gas distributor; New England Gas and Electric Association; Providence Gas Company; and several other smaller companies sharing a minority interest. For financing assistance, Poe had included First Boston Corporation.

The company's interest in New England was fortified by the interest expressed by New England companies and U.S. officials. In February 1948, Philip J. Philbin, a congressman from Massachusetts, sent a telegram to Texas Eastern indicating both an interest in the company's expansion plans and willingness to help the company achieve its expansion goals. Hargrove responded to Philbin that the company currently could not "serve more than a small segment of the area desiring service from us" even with an increase in the line's capacity. Hargrove reported that the company was preparing to file an application with the FPC to construct a third line to the Northeast and increase its capacity. He expressed confidence that the line would be built. Hargrove told Philbin "the development of such a project is no casual matter. It requires the assembling of large supplies of gas over a long period; the raising of tremendous sums of money through the issuance and sale of securities; and finally the securing of a very substantial tonnage of steel for fabricating the necessary pipe."[15] Although Hargrove did not specifically mention the need for FPC approval, the items he listed were the major points that the FPC required before granting a certificate.

Soon after responding to Philbin's inquiry, Texas Eastern made official in March 1948 its plans to serve New England. Initially, Texas Eastern planned to extend its existing system to Boston, Massachusetts, and construct lateral lines to points in Connecticut, Rhode Island, Massachusetts, and New Hampshire. This plan reflected Texas Eastern's intention to sell gas directly to the New England gas distributors with whom Holley Poe had previously negotiated. Texas Eastern then filed an application with the FPC to construct a third pipeline parallel to the Big and Little Big Inch pipelines. The third pipeline was to be 26 inches in diameter and extend from Longview, Texas, to a point near the town of Wind Ridge in southwestern Pennsylvania. It was to be a total of 1,020 miles long with 314,740 hp and a capacity of 425 mmcf/d. The estimated cost of the line was approximately $152,131,000.[16] The added capacity from the proposed pipeline would increase Texas Eastern's total capacity to approximately 933 mmcf/d. The plan provided for part of the line's 425 mmcf/d to be delivered to various customers in New England. Although the company filed the application with the intent to proceed, it did not pursue the plan aggressively because arrangements which supposedly had been made between Texas Eastern and certain New England utility companies fell apart at the last minute.[17]

In Boston to close the deal for the venture, Poe received a telephone call at his hotel before leaving to meet with the other partners in the deal and sign the papers. A representative of the financial institution that had tentatively agreed to finance the project told him that pressure had been brought on them by outside sources and they were backing out of the deal. Whether the pressure came from a competing company or from the other utilities is uncertain. But when the local companies involved came to believe that their share of the proposed profits might be insufficient, the deal broke down. One journal vividly described the situation: "This proposal got a freeze-out which is memorable even in New England, where the fiscal climate has long been adverse to outsiders who propose to make money New Englanders might be making themselves."[18] With the disruption in Texas Eastern's original plan to sell gas directly into the New England area, the company modified its plans to construct an additional 26-inch line and lowered its original capacity requirement. The prospects for the New England market remained bright, but the company's immediate plans to become its dominant supplier stalled.[19]

After Texas Eastern's initial unsuccessful attempt to sell gas into New England, the company entered negotiations to revitalize a version of the original deal. In 1948, Reginald Hargrove developed a new strategy. He proposed to form a jointly owned pipeline to be named the Algonquin Gas Transmission Company with the same New England distributors Poe had dealt with a year earlier. In this plan, Texas Eastern would not own a controlling share in the pipeline, but it would be Algonquin's major, if not sole, natural gas supplier. To supply Algonquin, Texas Eastern proposed to extend its system directly into New England. Texas Eastern also commenced plans to construct a third trunkline to increase its overall system capacity. Dubbed the Kosciusko line after its point of origin at Kosciusko, Mississippi, the new 30-inch line would receive 134 bcf per year from United Gas Corporation.[20]

Hargrove's new approach to Texas Eastern's plan to enter the New England market succeeded. On September 28, 1949, Algonquin Gas Transmission Company came into being. The participants in the deal designed Algonquin to connect Texas Eastern's lines with major New England utilities. These regional gas distributors, which in 1948 produced more than 50 percent of the manufactured gas used in Massachusetts, Rhode Island, and Connecticut, controlled the company while Texas Eastern had a minority interest. On January 24, 1950, Algonquin filed an application with the FPC to construct and operate a natural gas line connecting with Texas Eastern facilities in New Jersey and running north to the Boston area. Algonquin proposed to serve most sections of all New England states where economically feasible except Vermont and Maine.[21]

COMPETITIVE STRATEGY FOR CERTIFICATION

Gardiner Symonds wasted little time in responding to Texas Eastern's challenge by mapping his own strategy to enter the region. His company's original plan to enter New England faltered in April 1948 for similar reasons. After the FPC approved Tennessee's scaled-back application in

July 1949, certifying it to sell gas to Buffalo, New York, the company revitalized its New England expansion plan by proposing to extend its main line from Buffalo to the New Hampshire–Massachusetts state border, where it would connect with a new New England pipeline system to be controlled solely by Tennessee Gas. In August 1949, a few weeks before the incorporation of competitor Algonquin, Symonds organized the Northeastern Gas Transmission Company, a wholly owned subsidiary of Tennessee Gas Transmission, to compete with Algonquin for both regulatory certification and New England markets.[22]

Northeastern then neutralized any plans Transcontinental may have had to build its own line into New England by agreeing to purchase gas from it. On April 28, 1950, the FPC approved Transco's application to increase its gas sales to 505 mmcf/d. Sixty-five mmcf/d of the total would be dedicated to Northeastern; the South Jersey Gas Company would take delivery of the gas until Northeastern was ready to accept it. Transcontinental began plans to build its own New England extension to connect with Northeastern Gas at the New York–Connecticut border. In addition, Tennessee Gas proposed to sell Northeastern 120 mmcf/d through its Buffalo extension. Arranging gas supply quickly and acting fast to put Northeastern in motion gave Tennessee Gas an advantage over Texas Eastern in this phase of the competition. In a matter of weeks, two major companies put forward competing proposals to supply natural gas to Boston and the remainder of New England.[23]

Northeastern aggressively pushed the certification process to force the FPC to certify its New England pipeline. The company applied for a certificate on August 24, 1949, for the estimated $17 million New England expansion program. The pipeline would consist of 511 miles including laterals and have a capacity of 350 mmcf/d. Northeastern had already contracted for gas sales to distribution utilities in the six New England states. Other prospective customers were then considering contracting for gas. But some of the largest gas utilities in Massachusetts, particularly in Boston, were already committed instead to Algonquin.[24]

Symonds refused to accept competition from the proposed Algonquin line. Soon after organizing Northeastern, he traveled to Maine to present a speech, "Natural Gas for New England," to the New England Council. Reminding his audience of spring 1947, when he said Tennessee Gas would work to bring natural gas to New England, he now told them that his Northeastern subsidiary would sell gas to New Englanders purchased from both Tennessee Gas and Transcontinental. Symonds also reviewed his company's history and general expansion plans aimed at bringing natural gas to New England.

In his talk, Symonds took a not-so-discrete shot at Texas Eastern. Symonds emphasized that Northeastern would not be a promotional scheme. "There will be no promotion stock; no founders' share to divide among insiders. Tennessee has not and will not do business that way."[25] However, Symonds had to deal with the fact that Tennessee Gas would own all of Northeastern. Symonds noted that Alexander Macomber of Boston would be chairman of Northeastern's board and that several other New Englanders would be included as well. Others associated with Tennessee Gas included Daniel H. Morrisey of Rhode Island, who had been affiliated with Stone & Webster, which had a controlling stock interest in Tennessee. For legal counsel, Northeastern hired Thomas H. ("Tommy the Cork") Corcoran for representation at the FPC. Corcoran, a former New Deal legal aide to Franklin Roosevelt, had previously associated with one of the unsuccessful bidders for the Inch Lines.

Symonds justified Tennessee's ownership of Northeastern by stating that the parent company needed full control over its subsidiary, particularly during its formative years, in order to ensure that it operated correctly. Tennessee Gas would not own Northeastern permanently, he indicated, and New England investors would have the opportunity to participate in ownership, presumably through public stock offerings.[26]

The substantial differences between the organization of Northeastern and Algonquin influenced public opinion in New England. Despite Symonds's defense of Tennessee Gas's complete ownership of Northeastern and pledge that it would not always own it, his plan aroused considerable resentment of "outside capital." Actually, Tennessee Gas opposed local ownership of Northeastern on the grounds that the local utilities retained huge investments in manufactured-gas plants and were dominated by coal interests hostile to natural gas. Under this scenario, these firms might well attempt to stifle or choke off the development of natural gas in New England. Local media accounts decried Tennessee Gas's total ownership and management of Northeastern and cited the need for "keeping New England for New Englanders." True to character, Gardiner Symonds did little to blunt such criticism; instead, he fueled the controversy by proclaiming that "we had no intention of giving away part of (Northeastern) for the privilege of doing business in New England."[27]

In contrast, Texas Eastern's more culturally sensitive project included a leading role for local interests; two New England utility companies,

Eastern Gas and Fuel Associates and New England Gas and Associates, owned over 70 percent of Algonquin.[28] In all, Texas Eastern owned only 28 percent of the venture, a fact Reginald Hargrove put to good use before audiences in New England; he described Algonquin as "a joint venture — one-fourth ours, three-fourths New England's."[29] Hargrove had learned from Poe's experience with Minute Man that success in the New England market would necessitate substantial participation by New England partners. Wisely, he promoted this view, which was in sharp contrast to Tennessee Gas's seemingly callous strategy.

Apparently, Hargrove's plan to supply gas to a new regional distribution system appealed to at least some New England utilities. In explaining his company's reasoning behind joining the Algonquin plan, F. D. Campbell, president of the New England Gas and Electric Association, sent Texas Eastern vice president George Naff proposed wording from NEGEA's annual report: "Independent engineers confirmed our earlier opinion that New England gas utilities would secure better terms, greater reliability and a more adequate supply if Algonquin were to build a pipe line system to serve part or all of New England than if they were to individually contract with Northeastern Gas Transmission Company, a wholly-owned subsidiary of Tennessee Gas Transmission Company."[30]

Aside from the highly public debate between Tennessee Gas and Texas Eastern, other groups were organizing to oppose the efforts of both pipeline companies to expand their service into New England. In a report delivered to Texas Eastern, a research group noted that since the summer of 1949, about one year after Tennessee Gas announced the formation of Northeastern, oil marketers who served the majority of New England's residential heating customers had begun working through various trade associations to keep natural gas out of New England. Citing how the introduction of natural gas into Chicago and Detroit "had [placed] oil marketers right over a barrel," fuel oil dealers were facing an economic threat similar to that of the manufactured-gas industry.[31]

The contest for New England markets continued in the regulatory arena, the FPC, and the media. During early 1950, the FPC began separate hearings for the applications of Northeastern and Algonquin to serve the New England market. At the very least, witnesses expressed a great interest in purchasing stable natural gas supplies. Even Paul Dever, governor of Massachusetts, presented a paper appropriately titled "Our Need for Natural Gas in Massachusetts and New England." However, the hearings also became a showcase for the intense competition between the two subsidiaries. During a recess in the hearings in April and early May, Algonquin realized that it might be able to serve at least some markets in New England faster if it proposed to share part of the market with Northeastern. In accord with this change in strategy, Algonquin filed an amendment on May 1, 1950, to its original FPC certificate application and proposed to serve only Connecticut, Rhode Island, and eastern Massachusetts, including the Boston and Worcester metropolitan regions.

Initially, the FPC indicated that a single company, or supplier, should serve the entire New England market. However, the commission soon expressed doubts about the ability of either one of the companies to fulfill the area's long-range market demands. In order to compare and monitor each plan, the FPC consolidated the applications of both Northeastern and Algonquin. The consolidated hearings began in July 1950.

During these hearings, Reginald Hargrove made several speeches to New England business groups both to garner support and to reassure them that adequate gas supplies would always be present. In one instance, Hargrove received an opportunity to speak before the New England Council in September 1950 on its twenty-fifth anniversary. He pointed out that the projected capacity of Algonquin as supplied by his company could, if need be, fulfill the New England demand during all seasons if the company did not first receive FPC permission to serve only part of New England. Hargrove, appearing flexible and even-handed, noted that although natural gas would benefit residential and commercial customers, most northeastern industries might find industrial bituminous coal and fuel oil cheaper to use in the near future. He presented his company as one that was there to serve customers, not fight off competition. He told the audience that Texas Eastern would gladly serve all of New England or part of the area. "We stand ready," said Hargrove, "to undertake to serve the entire New England market ourselves if this appears to be the most desirable manner for such service to be rendered."[32]

The current series of FPC hearings ended in September. The FPC then issued its Opinion no. 201 on October 4, 1950, regarding the hearings. In this opinion, the FPC overtly criticized both companies. The FPC remarked that each of them proposed to serve only certain distribution companies and not the entire New England area. Both proposals contained unnecessary duplication and overlapping facilities. If either proposal was certificated, the other would become infeasible and many parts of New England would remain without natural gas.

The commission blasted both Algonquin and Northeastern for disregarding the needs of the New England fuel market. "It is clear," the FPC reported, "that these applicants in proposing to make natural gas available to New England—a highly industrialized area with extremely high fuel costs, and one of the few remaining areas in this country without natural gas—have placed their own selfish interests, and those with whom they are associated, above the best interests of the public."[33] The commissioners went on to say that neither project met their requirements for serving New England. In concluding its opinion, the FPC said it was prepared to reopen the hearings if necessary.

Algonquin interpreted the FPC's opinion as an indication that it desired both companies to devise a plan to serve New England jointly without overlapping facilities yet provide service to the entire region, a position consistent with the amendment to its original certificate application. On this basis, Algonquin telegraphed Northeastern on October 5, in accord with FPC Opinion no. 201, and suggested that the two companies devise a plan to divide the New England market between them.

Northeastern's blunt and uncooperative reply came on October 9 in the form of an additional proposal filed with the FPC requesting that in a separate hearing, Northeastern should receive a certificate to serve, as quickly as possible, the entire New England market. After learning of Northeastern's proposal through a company-issued press release, an Algonquin lawyer, John Rich, telegraphed Gardiner Symonds: "Such press release and proposal seem to constitute complete rejection of our proposals to you."[34]

Symonds was quick to reply. Later the same day he telegraphed Algonquin that the FPC had not in fact endorsed any plan to divide the New England market between the two companies. Instead, Symonds noted that his companies were willing to spend large amounts of money to bring natural gas to New England unless

> the ultimate merchandising company is to be throttled by the control or stalemate of representatives of coal and other competing fuels. . . . Tennessee is convinced that Algonquin is under the domination of persons whose primary interest is the protection of the coal industry. Our experiences with you in the past year have convinced us that such domination would not only work against the interests of the New England public and the New England gas-distributing companies but also against the interests of Tennessee itself as a seller of natural gas. . . . In the interests of the people of New England . . . the one essential is single management responsibility concerned only with selling natural gas.[35]

Symonds's charge that coal interests controlled Algonquin appeared correct but, in fact, was not. Eastern Gas and Fuel Associates and New England Gas & Electric Association (NEGEA), both of which were partners in Algonquin, did have manufactured-gas properties. But the NEGEA had been investigating the possibilities of using natural gas since the fall of 1946. Significantly, Eastern Gas and Fuel Associates, now a major stockholder in Algonquin, had only recently intervened in, and opposed, both Texas Eastern's and Transcontinental's original certificate hearings to sell natural gas in the Northeast. Eastern Gas and Fuel had once been owned by Koppers Company, which was a major coke, coal, and manufactured-gas enterprise. However, recognizing the superiority of natural gas, Eastern Gas and Fuel allied with Algonquin and encouraged its subsidiaries to purchase natural gas as well. As Texas Eastern's Reginald Hargrove pointed out, Algonquin—including one of its major stockholders, Eastern Gas and Fuel—had already invested considerable money in natural gas and "no company is going to throw $3½ million away for a front."[36]

Despite its earlier criticisms of both Algonquin and Northeastern, the FPC chose to force the two companies to share the New England market. Northeastern had applied for a certificate first, and the FPC granted Northeastern a certificate first, on November 8, 1950. But the FPC ruled that Gardiner Symonds's Northeastern could sell gas only in what amounted to approximately 54 percent of the New England region, with the remainder to be reserved for Algonquin. The FPC gave Symonds thirty days to accept the certificate, and on November 10, Northeastern officially accepted it.[37]

Gardiner Symonds, quick to capitalize on his prized certificate, made a forceful presentation to the Boston Security Analyst Society meeting in December 1950. Symonds defended Northeastern's claim to the entire New England area by reiterating that only Northeastern had a certificate to sell natural gas into the region. He then attempted to enflame the passions of the analysts: "And what is more significant than anything else, [we have] a burning desire to get natural gas to New England and get it here quick, and again, to sell it under a tariff that will encourage its use."[38]

By default, however, Texas Eastern's affiliate claimed slightly less than half the regional market, but its share included the largest city in the area, metropolitan Boston. In a separate decision a few months later, the FPC authorized Algonquin to begin constructing its pipeline.[39] Texas Eastern then quickly signed a gas supply contract with Algonquin on May 16,

1951. Five days later, Algonquin signed contracts with nine New England utilities, most of which were located in eastern Massachusetts and in Rhode Island.

Symonds, however, had no intention of abiding by the FPC ruling which divided New England. The following month, Northeastern filed for a certificate to serve two additional small distribution companies and "all of the towns and communities which Algonquin Gas Transmission Co. has applied to this Commission for authority to serve."[40] The FPC approved Northeastern's application except for the parts allowing it serve Algonquin's proposed market area.

Symonds next appealed to the FPC for a rehearing, but the agency dismissed his plea. Clearly annoyed at Northeastern's tactics, the FPC stated that "Northeastern was permitted to develop its case in full, including its opposition to Algonquin's application, and it cannot now thwart the expeditious disposition of its competitor's application by filing a second exclusive application." In unusually harsh language, the FPC harangued Tennessee for manipulating and hampering the regulatory process: "If such actions were permitted, hearings might continue *ad infinitum* and regulation which was designed to protect the consuming public would fail of its purpose and become a time-consuming process."[41] The FPC realized that it might ultimately receive blame for delaying the introduction of natural gas into New England. But FPC chairman Thomas C. Buchanan disagreed with the FPC's majority decision to reject Northeastern's application. He claimed that a rejection of Northeastern's application was tantamount to giving Algonquin an unfair advantage in New England.

Soon thereafter, Northeastern applied for a certificate to increase the size of pipe in its Northeastern system from 20 to 24 inches. Northeastern's primary reason for wanting to increase the size of its pipe was that there was not enough 20-inch pipe available while 24 inch was available. The FPC discovered, to its dismay and further annoyance at Gardiner Symonds, that the only reason 20-inch pipe was not available to Tennessee was that Symonds himself, now chairman of Northeastern, had personally ordered its pipe supplier to roll 24-inch pipe instead of 20 inch. Until that date, 20-inch pipe was available. Although disturbed at Symonds's actions, the FPC decided that Northeastern's need for larger diameter pipe was justified by market demand.[42] With a certificate and operational pipeline into some parts of the region, Northeastern began selling gas into portions of New England on September 28, 1951.[43]

Northeastern received permission to use the larger diameter pipe, but it remained unable to garner a larger share of the New England market. Symonds continued to plot Algonquin's downfall. Still smarting over the FPC's refusal to allow Northeastern to serve Algonquin's market area, Symonds took his case to the U.S. Third Circuit Court of Appeals, claiming that Northeastern had been denied due process by the FPC. The court agreed and set aside previous commission actions and remanded the entire case to the FPC. This negated previous FPC decisions establishing market shares and also nullified the permit for the construction of Algonquin's pipeline. Subsequent appeals by Algonquin and Texas Eastern to the U.S. Supreme Court via writ of certiorari were denied. The FPC then set new hearings to begin November 1952.[44]

As the regulatory agency and the courts put forward their opposing views of the proper procedures for defining competition through regulation, months and then years were sliding past, with little natural gas reaching New England. This delay was not only frustrating executives at Texas Eastern, Algonquin, Tennessee Gas, and Northeastern, it was causing their gas suppliers to question their wisdom in dedicating gas to ventures hopelessly tied up at the FPC. Transcontinental, which had agreed to sell 64 mmcf/d to Tennessee Gas and build a New England extension for that purpose, was now publicly questioning whether it would ever deliver gas to Tennessee. Transcontinental's 1951 annual report stated that owing to "unavoidable delays" for Northeastern and Transco, Transco had made no deliveries of gas to Northeastern by the contract date, and no arrangements had been made to extend the contract. On October 29, 1951, Transcontinental notified Northeastern that it had canceled the gas supply agreement "because neither Transcontinental nor Northeastern had built the necessary facilities by September 1, 1951.[45] Clearly, Transco's self-interest was in signing good contracts with the many customers along its pipeline route who would immediately take and pay for the gas.

PUBLIC EXPOSURE

The hiatus between hearings was filled with media reports in magazines such as *Life*, *Fortune*, *Time*, and *Forbes* outlining the competition between Algonquin and Texas Eastern versus Northeastern and Tennessee Gas. Gardiner Symonds's aggressive tactics attracted particular media attention. Press reports generally speculated that Symonds's intense interest in New England and his antagonism toward Algonquin were sparked by his loss of the Inch Lines to Texas Eastern. "That's a bunch of bunk," Symonds responded. He was not averse, though, to injecting personal feelings into business. Referring to Algonquin's attempt to serve New England at Northeastern's expense, Symonds said, "They delayed us for two years. They harassed us and kept us over the griddle and made all the trouble they could for us. I'm just vindictive enough to want to do the same thing to them."[46]

When the Federal Power Commission finally reopened hearings on the New England matter on October 31, 1952, public opinion within the region seemed to favor the original FPC proposal that the two companies, Algonquin and Northeastern, share the market. Some public officials, including Massachusetts governor Paul Dever, also favored the dual-supply approach. After Northeastern rejected the dual-supply proposal, though, Dever sought to mediate a compromise settlement but Gardiner Symonds turned a deaf ear. Dever responded by reaffirming his support for the dual-supply plan along with the backing of Massachusetts Congressman Henry Cabot Lodge and Senator Leverett Saltonstall. A month later, the state's governor-elect, Christian Herter, added his endorsement for the dual-supply approach.[47] This support no doubt sprang from the local concern that New England's natural gas consumption, shaped as it was by the harsh conditions of nasty winters, could well demand the capacities of more than one supplier. As this debate went back and forth, the FPC's new hearings on Algonquin's certificate application continued relentlessly, lasting from November 1952 through the following June.

By the late winter of 1951–52, Northeastern had completed its entire New England line after battling not only the FPC but landowners as well. During the subsequent construction phases of the pipelines, both companies encountered localized and regional opposition throughout New England. Often, landowners resisted the pipeline company's attempts to purchase easements on which to lay their pipeline. In some circumstances, the company would invoke the power of eminent domain. In both Massachusetts and Connecticut, legislatures had adopted processes through which a pipeline company could proceed with the eminent domain process. Farmers and other landowners, however, resented "these Texas boys in their cowboy boots and their big hats . . . chopping down hedges and clearing the right-of-ways" and went directly to their respective legislatures to oppose the pipeline plans.[48] Northeastern was able to overcome local opposition and soon completed construction of its line.

Algonquin's line suffered more from regulatory delay than farmers' opposition. It stood some 100 miles short of completion, stalled by the courts pending the FPC's decision. The delay cost a reported $32,000 daily in idle men and materials, and the project threatened to overrun its estimated $50 million cost by a considerable margin. Algonquin appealed to the FPC for an emergency certificate to get construction under way again, but the commission refused pending the rehearing process. Press reports speculated that if the FPC ultimately refused to approve Algonquin, the company "faced the gloomy prospect of trying to sell its premature pipeline for what it could get to the only possible bidder: its No. 1 enemy, Gardiner Symonds and Northeastern."[49] With this unlikely but frightening prospect in mind, Texas Eastern's 1952 annual report expressed concern to stockholders about the future of Algonquin, indicating that competing interests had convinced the FPC to conduct further hearings on the Algonquin matter. Although Texas Eastern had already invested $4 million in the Algonquin project, it reported that it did not "consider the extent of such exposure after tax credits to be particularly significant in relation to the size of the company's other assets and investments."[50]

Symonds pressed the attack. He and his supporters argued that the increased costs of Algonquin's construction eventually would be passed on to the consumer with higher rates. The prospects of higher gas rates, according to Symonds, were still another reason supporting a single New England supplier, his own completed Northeastern pipeline. Symonds also went to Washington for help in his cause. Early in the struggle he had employed the services of Washington attorney and lobbyist Thomas Corcoran. Corcoran did what he could to gain congressional support for Symonds. In the summer of 1950, a *Life* magazine article reported that Massachusetts congressman John McCormack, Democratic majority leader and a long-time friend of Corcoran, had "convened a meeting which has never been matched in FPC annuals. Four of the five commissioners and several of the staff attended. New England representatives and senators rawhided the commissioners for upward of two hours, telling them in effect to quit stalling and get gas into New England—or else. Corcoran, apparently overcome by good taste, did not attend." Later, McCormack tried to clarify his own intentions about the meeting and his overall efforts in the matter: "I was for natural gas—period," he said. "I was for the consumer. If the position I took worked to the benefit of one company or the other, that was not my responsibility."[51]

Such a meeting no doubt strengthened the case for Northeastern, which at that point was substantially ahead of Algonquin in having a completed pipeline delivering gas. The same source commented on Texas Eastern's lack of allies in Washington: "Meanwhile, if the Brown-Hargrove group and their New England partners had any friends in court, they kept well out of public view. Secretary of Labor Maurice Tobin, of the Boston Tobins, is the only Washington figure of any note who took a manifest hand in their behalf, and he says his efforts were confined to four or five introductions at the commission and elsewhere."[52]

Tennessee's influence also seemed to have penetrated the FPC itself. *Fortune* published an article stating that FPC chairman Thomas Buchanan favored Tennessee Gas throughout the New England episode; indeed, he cast the only dissenting vote when the FPC refused to allow Northeastern to serve Algonquin's service area. The article tried to show how Tennessee Gas attorney Corcoran, a personal friend of Buchanan, had close ties with Senator Francis Myers of Philadelphia, a long-time opponent of the natural gas Inch Lines who had sponsored Buchanan's appointment to the FPC.

On Algonquin's side, President John F. Rich, a Boston attorney and investor in charge of Algonquin's strategy, noted Tennessee's highly public political allies and their maneuvering. "After all," he said, "Corcoran has had a liberal education in that kind of thing. I think we were less experienced and didn't do so much."[53] The press reported, however, that Governors Dever, Saltonstall, and Lodge urged the FPC to certify Algonquin.[54] The public nature of this highly charged political competition may have in fact forced the FPC to treat both companies fairly in order to deflect further criticism of itself.

The ultimate solution to the competition could come only when the FPC completed its hearings and made a decision. The basic problem continuing to face the FPC was its attempt either to choose one or the other pipeline to serve all of New England or to somehow divide the market between the two in a mutually acceptable manner. At the same time, potential natural gas consumers in Massachusetts were publicly asking the FPC to allow gas to flow first and "iron out [the] difficulties" later.[55] Supporting this view was Massachusetts governor-elect Herter. He told the FPC: "I have felt very strongly that the interest of the State of Massachusetts will be best served by having two separate companies serving the state, rather than having one single company operating as a single unit within the state as a whole." During the next month, Governor Lodge of Connecticut invited representatives of both Algonquin and Northeastern to meet with him to discuss ways of getting gas to his state that winter.[56]

By early 1953, however, the competition actually increased. Still on the offensive, Northeastern sought to attract public support to its position through publication of a series of full-page advertisements in the *Boston Herald*. One ad, titled "Who Built the Natural Gas Roadblock? Why was it Built? Will it be Removed?" attacked Eastern Gas and Fuel Associates along with Algonquin.[57] Within a few days, a *Boston Herald* editorial defended Algonquin's position.[58]

Another problem was the sheer magnitude of the task: marshaling information; filing legal briefs; presenting witnesses from both companies and their affiliates, retail distributors, and representatives of state and local governments; and listening to engineering specialists. The hearings involved eighty-three daily sessions and the appearances of 125 witnesses. The second round of FPC hearings took roughly the same amount of time as had been required to construct the Little Big Inch Pipeline. The war had prompted action; the creation of a new postwar regulatory regime in the natural gas industry at times seemed to paralyze all participants.

CORPORATE AND REGULATORY AGREEMENT

As the hearings plodded on, the FPC continued to seek a compromise similar to its initial ruling to divide the New England market between Northeastern and Algonquin. As months passed, Symonds and Hargrove, tiring of the costly delays, also seemed to moderate their initial positions. Appearing before the commission in March 1953, Symonds testified that even if Northeastern received the entire New England market, it would buy gas from both his company, Tennessee Gas, and its rival, Hargrove's Texas Eastern. Shortly after, in a shift from his earlier stance, Hargrove stated that recent surveys indicated substantial market growth in New England which would "rapidly outdistance the ability of . . . one [company] service to be adequate for New England."[59] Both parties subsequently agreed to conclude the hearings on June 3, 1953, present final closing arguments within fifteen days of that date, and request a final decision from the FPC by July 1, 1953. On July 1, Tennessee, Northeastern, Texas Eastern, Algonquin, and two distribution companies filed with the FPC a proposed settlement letter providing a dual natural gas supply for New England which the FPC accepted.[60] This conclusion to the long and drawn-out legal battle brought a degree of compromise and

New England pipelines. Dashed line indicates Northeastern Gas Transmission Company. Solid line indicates Algonquin Gas Transmission Company.

moderation, or at least a form of battle fatigue, which encouraged all participants to leave the hearing room and return to the business of selling gas. The settlement largely reaffirmed the commission's original decision. The two suppliers, Northeastern and Algonquin, shared the New England market with the latter retaining the Boston area (see map). In compensation for its exclusion from the Boston area, Tennessee Gas obtained the right to serve two local distribution companies previously served by Texas Eastern in Louisville, Kentucky, and Pittsburgh, Pennsylvania, as well as several small towns in Tennessee.[61] Just as the end seemed in sight, Gardiner Symonds introduced a new demand that threatened to stall the completion of a settlement. Previously, his company had filed an application with the FPC to supply gas to Niagara Gas Transmission Limited for Canadian markets. He subsequently announced that his acceptance of the New England settlement was conditional upon the commission's certification of an export license to fulfill the Canadian contract. At that point, several Canadian gas distribution companies scheduled to receive gas under the proposed Tennessee-Niagara contract intervened. They feared that if Symonds did not get what he wanted in New England, he would pull out of the Canadian supply deal, leaving Canadians without gas. As one intervenor put it, "Tennessee is thereby repeating its pattern of past conduct. It desires everything or nothing."[62] Symonds subsequently appeared ready to justify this charge. When the FPC delayed approval of Tennessee's export license and set further hearings on the matter, Symonds was reported to be in "no mood to carry out the original agreement [the Algonquin-Northeastern settlement] until his Canadian market is guaranteed."[63]

Within a few weeks the FPC granted Tennessee's export permit and Symonds accepted the New England settlement.[64] This strategy was a choice example of the intricacies of competition in a regulated environment. Symonds correctly judged the desire of the FPC to wrap up the New England controversy and avoid becoming embroiled in what promised to be a politically charged dispute over Canadian gas supplies. By bringing the two issues together, he maximized his overall gains while forestalling another lengthy regulatory battle on the Canadian issue.

In early July, the companies officially submitted their settlement to the FPC. The following month, on August 6, the FPC granted certificates to both Algonquin and Northeastern that reflected the FPC's earlier decision to essentially divide the market between the two. Concurrently, the FPC authorized Algonquin to finish construction of its pipeline. The FPC had suspended work on the pipeline as it neared completion, but Algonquin needed only one month to finish the system. Algonquin began delivering gas to its New England customers on September 2, 1953. The introduction of natural gas into New England by both Algonquin and Northeastern meant the end of the manufactured-gas industry's monopoly in northeastern gas sales (see table 7.1).

Ironically, the intensity and the length of the battle for the New England market were not reflected in high profits once the ventures were under

Table 7.1. Gas Sales of Utilities in New England, 1945–1959
(millions of therms)

Year	Natural Gas	Manufactured Gas
1945	0	321
1947	0	364
1949	0	375
1951	15	414
1953	214	220
1955	441	17
1957	625	16
1959	894	16

NOTES: New England includes Connecticut, Maine, Massachusetts, New Hampshire, Rhode Island, and Vermont.

A therm is equivalent to 100,000 Btu.

SOURCE: American Gas Association, *Historical Statistics of the Gas Industry* (Arlington, VA: AGA, 1964).

way. Perhaps profits had never been the primary prize at stake in the race to New England, which, after all, was not a market to rival the larger, more urbanized regions around New York City and Philadelphia. Much individual and corporate pride was on the line in this dramatic episode, which marked the first major confrontation of two aggressive competitors, each of which viewed itself as the leader in cross-country gas transmission.

Now that the FPC had dictated the pipeline companies' respective market areas, the high level of industrial and political competition was over. Northeastern and Algonquin had become parts of a regional natural gas supply and distribution network with only one purpose: to supply natural gas to the consumer. Tennessee Gas, which had pretended to include New England's interests in its Northeastern system, soon dissolved Northeastern Gas as a separate entity and took over its operation. The reason for this action was that Northeastern was not financially healthy; by merging with Tennessee Gas, Northeastern could delay the imposition of higher gas sales rates and improve its competitive position. To improve the overall operating characteristics of their respective systems, both Tennessee and Algonquin agreed in early 1954 to interconnect their lines at Southington, Connecticut.[65] The systems were interconnected and could supply each other's customers with gas, and the FPC, recognizing the oligopolistic nature of the pipeline industry, felt more comfortable with two separate firms serving the region rather than a single regulated monopoly.

The New England episode represented an important case study of regulated competition. Both Tennessee Gas and Texas Eastern, unrestrained by the FPC, would certainly have constructed their facilities to supply gas to New England sooner. However, in view of the history of unregulated competition between highly capital-intensive operations based upon distant natural gas reserves, unregulated expansion in the 1940s and 1950s might well have resulted in gas supply problems and inadequate pipeline systems connected only to the largest regional gas distributors. The FPC acted to ensure a stable gas supply for the maximum number of possible customers by certifying projects capable of operating effectively and efficiently. By dividing the New England market into two areas and certifying two competing companies to serve them, the FPC effectively regulated the expansion of the natural gas industry into New England by ensuring a secure supply from stable suppliers.

The New England episode was the last in a series of highly politicized gas pipeline expansions into the postwar Northeast. Perhaps more so than in either the Philadelphia or New York City cases, the competition for New England characterized most clearly the FPC's regulatory practice. The interstate gas pipeline industry would be an interconnected system of nonduplicative, capital-intensive pipeline companies. Although single customers initially had access to gas directly from only one supplier, the various regional pipeline systems were interconnected, allowing gas to flow, for either normal business or emergency reasons, from producer to consumer.

8. The End of the Entrepreneurial Era

NINETEEN FIFTY-FOUR marked the end of the major episode of natural gas industry expansion from the Southwest to the Northeast and the beginning of a new era of producer price regulation heralded by the Supreme Court's Phillips decision. From this year forward, the FPC began a long, tortuous, and unsuccessful task of determining just and reasonable prices for gas produced for interstate commerce. After seventeen more years of various regulatory attempts to determine fair prices, the FPC's primary accomplishment was keeping the price of interstate gas so low that producers began selling less gas to the interstate pipelines and more to the intrastate lines, which, unregulated by federal authorities, could pay more for natural gas. Thus, by the mid-1970s, northeastern and other gas importing states were suffering from gas shortages while consumers in producing states continued in most cases to be insulated from shortages. The gas crises of the 1970s, along with the broader energy crisis, received a great deal of attention from scholars and energy writers. The era of abundant supply transported by new pipelines remained unstudied.

Certainly, there are good reasons to study the breakdown of market systems, unregulated and regulated. It is obviously important to understand how the market failed to implement preventive remedies for the future. It must be equally important to understand the successful operation of market systems, regulated or not. The controversy over federal producer price regulation can be traced back to the congressional hearings over the Natural Gas Act of 1938. The crucial issue of producer price regulation, however, has overshadowed any other lesson to be learned from the regulated gas industry.

Between 1938 and 1954, issues of regulated competition for markets, fuel conversion, and the dynamics of rapid post–World War II industry growth under what amounted to a New Deal era regulatory system characterized the natural gas industry. The competition for markets, the competition for the ability to actually enter the pipeline industry and compete for those markets, occurred for the most part in certificate hearings held before the FPC. The efforts expended at acquiring a certificate represented attempts, not always successful, to enter an industry. This era provides an interesting example of the process of entry regulation during a period of rapid industrial growth.

The FPC successfully facilitated the basic operation of market forces through the certification process. Applicants attempted to prove that they had adequate gas supply, consumer demand, engineering competence, and ability to finance the construction of the proposed line. These requirements reflected problems in the early gas pipeline industry of inadequate gas supply, primitive engineering technology, and uncertain financing — which often left the consumer without fuel. Nevertheless, the certificate was a standard for verification of ability to operate successfully in a highly capital-intensive industry, and it was normally gained only after lengthy public hearings that allowed various interest groups to participate in the certification process.

This pluralistic method of certificate-style regulation of gas pipelines worked during a period of rapid growth, abundant supply, and low prices. The difficult issue of price regulation remains less important in this study precisely because between 1938 and 1954 inflation was nominal, prices were low, and transportation capacity was increasing rapidly. The economic function of supply and demand worked without overt hindrance from destructive forms of regulation. Pipelines successfully delivered southwestern natural gas to Appalachia, the Philadelphia area, metropolitan New York City, and New England. Except for Appalachia, all these regions had natural gas for the first time, and they required increasingly more gas on a regular basis. At the same time, the manufactured-gas industry, characterized by increasingly expensive and dirty manufactured-gas plants, was on the decline.

Changes in gas market structure and regulatory rules after 1954 forced pipeline companies to adapt to new business conditions. The large northeastern gas markets were absorbing natural gas, and industry observers foresaw that a new corporate strategy would have to replace the entrepreneurial strategy which brought natural gas into the Northeast.

Soon after the completion of Transcontinental's line, President Claude Williams announced the end of rapid expansion in the natural gas industry: "There will be no more transcontinental projects; . . . all future expansion will be the result of additions to existing systems with the possible exception of a line going to the Pacific Northwest."[1] Gardiner Symonds made a similar observation when he remarked that "the race for franchise territory is about over. Pretty near everybody is served with gas."[2] Texas Eastern's Reginald Hargrove echoed this sentiment best when he summed up the achievements of the industry in a speech to a group of investment bankers in the early 1950s. After surveying the expansion of gas service to new markets in these years, Hargrove observed: "If the era of territorial expansion of our industry is coming to an end, that of developing the potentialities of our present areas has scarce begun."[3]

None of the entrepreneurs who originated the idea and early organization of each of the three pipelines, except for Gardiner Symonds, remained for long in the firm he had formed. Curtis Dall, forced out of Tennessee Gas after the Chicago Corporation took control of it, attempted on several occasions to sell natural gas production packages to Tennessee Gas and to otherwise profit from the firm's success through legal action. E. Holley Poe, reduced to the status of lower level board member after the formation of Texas Eastern, died in 1951 without a single mention in the firm's annual report. Reginald Hargrove, also a founding member of the firm and its first operational president, died in a plane crash in 1954 which took the lives of several other original stockholders of Texas Eastern as well as Thomas Braniff, founder of Braniff Airlines. Claude Williams fared poorly at Transcontinental and after Stone & Webster purchased a controlling interest in the firm in 1953, he was forced out of any active role in the company he created.

The end of the entrepreneurial era brought forth not only a new phase in regulatory policy but new directions for the regulated firms. For Texas Eastern and Tennessee Gas in particular, developing current market areas was not enough. Following trends in the American economy, these companies began to diversify their operations beyond natural gas transmission. In a sign of approval, the investment community applauded the move toward the development of new businesses. August Belmont, Texas Eastern's banker, recalled that "we did see a very positive move . . . to employ their resources in businesses other than the pipeline business."[4] While Belmont commented upon Texas Eastern's condition, he expressed

Table 8.1. Sales of Natural Gas in the Northeast
(bcf)

Year	Tennessee Gas	Pipeline Gas Sales Texas Eastern	Transcontinental Gas	Utility NG Sales in Northeast	Percent of total by 3
1945	74	—	—	575	13
1946	95	—	—	616	15
1947	109	39	—	705	21
1948	155	118	—	801	34
1949	221	162	—	854	45
1950	386	291	—	1,126	60
1951	386	339	136	1,335	65
1952	454	371	183	1,507	67
1953	496	418	192	1,623	68
1954	506	430	198	1,788	63

Sales do not include direct sales by pipelines for industrial use.

NOTES: This table is meant to suggest the magnitude of gas sales by the three pipelines to northeastern utilities but is not an inclusive accounting of all gas sales to, or by, northeastern utilities.

The Northeast includes all states within the possible significant service area of the three pipelines. These states include all of New England, New York, New Jersey, Pennsylvania, Ohio, Wisconsin, Michigan, Illinois, Indiana, Delaware, Washington, DC, Georgia, Maryland, North Carolina, South Carolina, Virginia, and West Virginia.

SOURCES: John Sherman Porter, ed., *Moody's Manual of Investments: American and Foreign* (New York: Moody's Investors Service, 1946–55) and American Gas Association, *Historical Statistics of the Gas Industry* (Arlington, VA: AGA, 1964.)

the investment community's consideration of the gas pipeline business as a whole. Both Texas Eastern and Tennessee Gas shared these sentiments and investigated additional business. Tennessee Gas would become one of the nation's largest conglomerates, owning large oil and gas production facilities as well as shipbuilding and farm equipment firms. Texas Eastern became one of the original players—although never a major one—in North Sea gas and oil production, the largest retail LPG distributor in the nation, and a major owner and developer of downtown Houston real estate. Transcontinental resisted diversification more than Tennessee Gas and Texas Eastern and remained more focused on its northeastern markets, but its strategy was the exception rather than the rule. These three pipelines, however, continued to be the primary suppliers of the natural gas used in the Northeast since the 1950s (see table 8.1).

During the period 1938 to 1954, the Federal Power Commission controlled the intense market-driven growth of the natural gas industry. This episode of regulated enterprise deserves recognition as a prime model of regulatory management of industrial growth. It suggests that pluralism, the interaction of various business and governmental interests in a forum designed to reflect market forces, can operate with reasonable success. Although the regulation of mature and stable economic systems may well present different problems than those of young and expanding ones, regulation designed to carefully facilitate, rather than overtly manipulate and distort, operative market forces can perform in the broadly defined public interest.

Notes

CHAPTER 1

1. This thesis generally agrees with Stephen Breyer's "normative approach":
    "It assumes that regulations seek in good faith to attain such goals [reason-
    able regulatory goals enacted by reasonable human beings], regardless of
    the existence of other possible motives in fact." Stephen Breyer, *Regulation
    and its Reform* (Cambridge: Harvard University Press, 1982), 10.
2. Bruce M. Owen and Ronald Braeutigam, *The Regulation Game: Strategic
    Use of the Administrative Process* (Cambridge, MA: Ballinger Publishing,
    1978), 68.
3. M. Elizabeth Sanders, *The Regulation of Natural Gas: Policy and Politics,
    1938–1978* (Philadelphia: Temple University Press, 1981), 23.
4. Phillips Petroleum Company v. State of Wisconsin, 347 U.S. 672 and 74
    Sup. Ct. 794.
5. See Richard H. K. Vietor, *Energy Policy in America since 1945: A Study of
    Business-Government Relations* (Cambridge: Cambridge University Press,
    1984); Sanders, *Regulation of Natural Gas*; and Arlon R. Tussing and
    Connie C. Barlow, *The Natural Gas Industry: Evolution, Structure, and
    Economics* (Cambridge, MA: Ballinger Publishing, 1984).
6. Ibid.
7. Each of these company names changed slightly during the period of this study.
    Except when the company names appear in quotations, they will be referred to
    as Tennessee Gas, Texas Eastern, and Transcontinental throughout the text.
8. See Thomas McCraw, *The Essential Chandler* (Boston: Harvard Business
    School Press, 1988), 361: Chandler defines an entrepreneur as one who
    owns or controls large shares of stocks and has "a major say in the selection
    of managers, in the making of over-all policy, and in long-term planning and
    allocation of resources."
9. During the 1970s, these and other pipelines curtailed gas sales to their
    northeastern customers, but this was due generally to a failure in regulatory
    policy toward gas producers, not a failure of pipeline regulation.

10. Thomas K. McCraw, "Regulation in America: A Review Article," *Business History Review*, 44, no. 2 (Summer 1975), 159–83. For a survey of relevant regulatory literature see Richard Vietor, *Energy Policy in America since 1945*, 8. Vietor discussed "public interest," termed "capture" "private interest," and added a third category characterized by bureaucratic and behavioral modes of analyses.

11. Ralph K. Huitt, "Federal Regulation of the Uses of Natural Gas," *American Political Science Review*, 46, no. 2 (June 1952), 455.

12. For a review of recent literature on entry regulation in the food and advertising industries see Robert L. Wills, Julie A. Caswell, and John D. Culbertson, *Issues after a Century of Federal Competition Policy* (Lexington, MA: Lexington Books, 1987), 185–234.

13. Joseph P. Kalt, "Market Power and the Possibilities for Competition," in Joseph P. Kalt, ed., *Drawing the Line on Natural Gas Regulation* (New York: Quorum Books, 1987), 115.

14. In Michael W. Klass and William G. Shepherd, eds., *Regulation and Entry: Energy, Communications, and Banking* (East Lansing: MSU Public Utilities Papers, 1976), 70–71.

15. Alfred E. Kahn, *The Economics of Regulation: Principles and Institutions*, Vol. 2 (New York: John Wiley & Sons, 1970), 5:2:152–53.

16. Stephen G. Breyer and Paul W. MacAvoy, *Energy Regulation by the Federal Power Commission* (Washington, DC: Brookings Institution, 1974), 5.

17. Paul W. MacAvoy, *The Regulated Industries and the Economy* (New York: W. W. Norton, 1979), 15–17; Robert W. King, "Building an Institutionalist Theory of Regulation," *Journal of Economic Issues*, 22, no. 1 (March 1988), 197.

18. Sanders, *Regulation of Natural Gas*, 22.

19. Thomas K. McCraw, *Prophets of Regulation* (Cambridge: Belknap Press, 1984), 305.

20. Alfred D. Chandler, Jr., ed., *The Railroads: The Nation's First Big Business* (New York: Harcourt, Brace & World, 1965), 185.

21. Hepburn Act (1906): 34 U.S. Stat. 584. Also see Donald R. Whitnah, *Government Agencies* (Westport: Greenwood Press, 1983), 291–96. The Hepburn Act of 1906 increased the ICC's regulation over railroads by prescribing railroad accounting and recordkeeping practices, gave the commission the power to replace unreasonable rates with rates determined by the commission, and placed additional forms of railroad commerce under ICC jurisdiction. See Arthur M. Johnson, *Petroleum Pipelines and Public Policy, 1906–1959*, (Cambridge: Harvard University Press, 1967), 23–32.

22. See Hepburn Act (1906): 34 U.S. Stat. 584.

23. Johnson, *Petroleum Pipelines and Public Policy, 1906–1959*, 206, 395.

24. Ibid., 189. Also see Transportation Act (1920): 41 Stat. 456.

25. See William R. Childs, *Trucking and the Public Interest: The Emergence of Federal Regulation, 1914–1940* (Knoxville: University of Tennessee Press, 1985).

26. John Richard Felton and Dale G. Anderson, *Regulation and Deregulation of the Motor Carrier Industry* (Ames: Iowa State University Press, 1989), 4, 5.
27. Kahn, *Economics of Regulation*, 5:2:14.
28. Felton and Anderson, *Regulation and Deregulation of the Motor Carrier Industry*, 12.
29. Kahn, *Economics of Regulation*, 5:2:14.
30. Motor Carrier Act (1935): 49 Stat. 543.
31. Breyer, *Regulation and its Reform*, 226.
32. Felton and Anderson, *Regulation and Deregulation of the Motor Carrier Industry*, 19.
33. Kalt, "Market Power and the Possibilities for Competition," 115.
34. Federal Power Act (1935): 49 Stat. 838–51.
35. Civil Aeronautics Act (1938): 52 Stat. 987.
36. Breyer, *Regulation and its Reform*, 200, 199.
37. Ibid., 226, 205.
38. Transportation Act (1940): 54 Stat. 899.

CHAPTER 2

1. Louis Stotz and Alexander Jamison, *History of the Gas Industry* (New York: Stettiner Bros., 1938), 78–79; idem, "Natural Gas," *Fortune* (August 1940); Leon F. Terry, "Natural Gas for the Northeastern Seaboard," *Mining & Metallurgy* (July 1947). Also, for a useful chronology of the gas industry see Dean Hale and Pat Loar, "Diary of an Industry," *American Gas Journal* (October 1959), 21–53.
2. Stotz and Jamison, *History of the Gas Industry*, 298.
3. For a good review of the European origins of the gas industry see Trevor I. Williams, *A History of the British Gas Industry* (Oxford: Oxford University Press, 1981).
4. American Gas Association, *Gas Rate Fundamentals* (Arlington, VA: AGA, 1987), 2.
5. Ibid.
6. Ibid.
7. Ibid., 3.
8. Btu refers to British thermal unit. One Btu equals the amount of energy required to raise the temperature of one pound of water by one degree fahrenheit.
9. See "Price Current of the Gas-Light Shares of Cities and Towns," *American Gas-Light Journal*, no. 1 (July 1, 1859), 2–3.
10. American Gas Association, *Gas Rate Fundamentals*, 6–7.
11. Martin V. Melosi, *Coping with Abundance: Energy and Environment in Industrial America* (New York: Alfred A. Knopf, 1985), 38–39.
12. Alfred M. Leeston, John A. Crichton, and John C. Jacobs, *The Dynamics of the Natural Gas Industry* (Norman: University of Oklahoma Press, 1963), 5–6.

13. Stotz and Jamison, *History of the Gas Industry*, 78–79.
14. Tussing and Barlow, *Natural Gas Industry*, 34.
15. American Gas Association, *Gas Rate Fundamentals*, 12–13; Stotz and Jamison, *History of the Gas Industry*, 80–81.
16. Tussing and Barlow, *Natural Gas Industry*, 33.
17. Robert W. Gilmer, "The History of Natural Gas Pipelines in the Southwest," *Texas Business Review* (May–June 1981), 130–31. Also see Ralph E. Davis, "Natural Gas Pipe Line Development During the Past Ten Years" *Natural Gas*, 16, no. 12 (December 1935), 3.
18. American Gas Association, *Gas Rate Fundamentals*, 14.
19. Sanders, *Regulation of Natural Gas*, 25.
20. American Gas Association, *Gas Rate Fundamentals*, 13.
21. Joseph A. Kornfeld, "Natural Gas: Newest of the Energy Giants," *Oil and Gas Journal* (May 1951), 449.
22. Gilmer, "The History of Natural Gas Pipelines in the Southwest," 131–32; Stotz and Jamison, *History of the Gas Industry*, 309–89.
23. Federal Trade Commission, *Report to the Senate on Public Utility Corporations*, Senate Document no. 92, 70th Cong., 1st sess. (summary and recommendations in pt. 84-A), 593–94, hereafter referred to as FTC Report.
24. Stotz and Jamison, *History of the Gas Industry*, 381–3.
25. Tussing and Barlow, *Natural Gas Industry*, 39.
26. Stotz and Jamison, *History of the Gas Industry*, 384.
27. Stotz and Jamison, *History of the Gas Industry*, 342–47, 362–65; Tussing and Barlow, *Natural Gas Industry*, 32.
28. Tussing and Barlow, *Natural Gas Industry*, 33–40.
29. Stotz and Jamison, *History of the Gas Industry*, 214–18.
30. Ibid.
31. Ibid.
32. American Gas Association, *Gas Rate Fundamentals*, 76–77.
33. Ibid., 14.
34. Sanders, *Regulation of Natural Gas*, 24.
35. See Carl D. Thompson, *Confessions of the Power Trust* (New York: E. P. Dutton, 1932).
36. FTC Report, pt. 84-A (1935). Also see William E. Leuchtenburg, *The Perils of Prosperity, 1914–32* (Chicago: University of Chicago Press, 1958), 190–91.
37. Also see Federal Trade Commission, *Investigation of Concentration of Economic Power, no. 36*, published as Temporary National Economic Committee Report no. 76-3 (Washington, DC: GPO, 1940), hereafter referred to FTC Monograph no. 36.
38. FTC Report, 73-A, 75.
39. Ibid., pt. 84-A, 87–95.
40. Ibid., 615–16, 609.

41. Ibid., 593.
42. Ibid., 591.
43. Kornfeld, "Natural Gas: Newest of Oil Giants," (May 1951), 432–50.
44. Francis X. Welch, "Functions of the Federal Power Commission in Relation to the Securities and Exchange Commission," *George Washington Law Review* 46, no. 4 (1956), 81.
45. Ellis W. Hawley, *The New Deal and the Problem of Monopoly* (Princeton: Princeton University Press, 1966), 336.
46. Richard W. Hooley, *Financing the Natural Gas Industry: The Role of Life Insurance Investment Policies* (New York: Columbia University Press, 1961), 45.
47. Dozier A. DeVane, "Highlights of the Legislative History of the Federal Power Act of 1935 and the Natural Gas Act of 1938," *George Washington Law Review* 14, no. 1 (December 1945), 34.
48. "Memorandum on Water Power Legislation," October 25, 1917, File Record Group-138, FPC, History of the Federal Power Act, 13–14, National Archives; Record Group hereafter referred to as RG; National Archives hereafter referred to as NARA.
49. Gifford Pinchot to Scott Ferris, September 17, 1918, RG-138, History of the Federal Water Power Act, 804, NARA.
50. 52 U.S. Stat. 821.
51. Vietor, *Energy Policy in America since 1945*, 70.
52. *Congressional Record*, 75th Cong., 1st sess., vol. 81, pt. 8, August 19, 1937, 9312.
53. Ibid., 9316.
54. DeVane, "Highlights of the Legislative History of the Federal Power Act of 1935 and the Natural Gas Act of 1938," 38–39. Also see U.S. Congress, House, Committee on Interstate and Foreign Commerce, Hearings on Natural Gas, 74th Cong., 2nd sess., April 1936, 156–58 (hereafter referred to as H.R. IFC 1936), and Gerald R. Nash, *United States Oil Policy, 1890–1964* (Westport: Greenwood Press, 1976), 214.
55. Sanders, *Regulation of Natural Gas*, 66, 67.
56. John G. Clark, *Energy and the Federal Government: Fossil Fuel Policies, 1900–1946* (Urbana: University of Illinois Press, 1987), 280.
57. H.R. IFC 1936, 156–58. Also see U.S. Congress, House, Committee on Interstate and Foreign Commerce, Hearings on the Natural Gas Act, 75th Cong., 1st sess., March 1937, 122, 134, hereafter referred to as H.R. IFC 1937.
58. Sanders, *Regulation of Natural Gas*, 49.
59. DeVane, "Highlights of the Legislative History of the Federal Power Act of 1935 and the Natural Gas Act of 1938," 39–41.
60. 52 U.S. Stat. 821, 1938.
61. Sanders, *Regulation of Natural Gas*, 50.
62. 52 U.S. Stat. 825.

63. *Congressional Record*, 75th Cong., 1st sess., vol. 81, pt. 6, April 22, 1937, 3771.
64. Vietor, *Energy Policy in America since 1945*, 70–72.
65. H.R. IFC 1937, 134. Also see Sanders, *Regulation of Natural Gas*, 49–50.
66. Clyde L. Seavey, "Federal Regulation of the Transportation and Sale of Natural Gas in Interstate Commerce," *American Gas Association Proceedings* (New York: AGA, 1938), 126–29.
67. "Natural Gas," 56.
68. Ibid., 96.
69. Ibid., 99.

CHAPTER 3

1. John W. Frey and H. Chandler Ide, *A History of the Petroleum Administration for War, 1941–1945* (Washington, DC: GPO, 1946), 227.
2. D. P. Hartson, "The Appalachian," *Proceedings of the American Gas Association* (1941), 132.
3. Frey and Ide, *History of the PAW*, 229.
4. Ibid.
5. Ibid., 228–29.
6. J. French Robinson, "Demand and Supply of Natural Gas after the War," *AGA Procedings—1942* (New York: AGA, 1942), 30.
7. Harold L. Ickes, *Fightin' Oil* (New York: Alfred A. Knopf, 1943), vii.
8. J. R. Parten, interview by Christopher J. Castaneda and John King, June 27, 1988, Texas Eastern Corp. History Collection.
9. Bill Murray, interview by Christopher J. Castaneda, December 12, 1990, Oral History of the Houston Economy, University of Houston.
10. ALCOA to the Office of Production Management, November 14, 1941, RG-253, box 2924, "Letters from _____ to L. C. Tonkin in Support," NARA.
11. "Memorandum of Minutes of Conference on Natural Gas Supply in Appalachian Area, RG 253, box 294, "Hope Natural Gas Co., Cornwell, W. VA, to Perryvill, LA, original application," NARA.
12. E. Holley Poe, "Our Joint War Job," *Proceedings of the American Gas Association* (1942), 61.
13. See J. French Robinson, "Demand and Supply of Natural Gas after the War," *Proceedings of the American Gas Association* (1941), 29–33. Also see Grover T. Owens, "The Responsibility of the Natural Gas Industry in Post-War Planning," *Proceedings of the American Gas Association* (1942), 79–84.
14. Frey and Ide, *History of the PAW*, 230.
15. J. A. Krug to J. S. Knowlson, no date, in folder "Natural Gas-L-31," RG-179, box no. 6, "L" orders, NARA.

16. Memorandum by General H. K. Rutherford, January 13, 1942, "Natural Gas—L-31," RG-179, WPB "L" orders, box no. 6, NARA.

17. "Limitation Order L-31," RG-179, box no. 6, "Natural Gas—L-31," NARA.

18. Ernest R. Acker, "The Gas Industry in War," *Proceedings of the American Gas Association* (1942).

19. Frey and Ide, *History of the PAW*, 229.

20. Kenneth S. Davis, *FDR: The Beckoning of Destiny, 1882–1928: A History* (New York: G. P. Putnam's Sons, 1971), 799; idem, *FDR: The New Deal Years, 1933–1937: A History* (New York: Random House, 1986), 165–66, 341, 618.

21. Curtis B. Dall, *FDR: My Exploited Father-In-Law* (Tulsa: Christian Crusade Publications, 1967), 123.

22. In 1946, the company's name changed slightly to Tennessee Gas Transmission Company.

23. Tennessee Gas and Transmission Company, *Minutes*, May 4, 1940, Tenneco, Inc., History Collection.

24. Curtis B. Dall to E. I. Dupont deNemours and Company, June 7, 1940, ibid.

25. Tennessee Gas and Transmission Company, *Minutes*, August 13, 1940, ibid.

26. John E. Buckingham to Curtis B. Dall, June 10, 1940, ibid.

27. Buckingham to Dall, December 10, 1940, ibid.

28. Buckingham to Dall, August 6, 1940, ibid.

29. "Re: Tennessee Gas & Transmission Co.," *Public Utilities Reports*, docket 2506, vol. 40 (September 11, 1941), 131.

30. U.S. Congress, House, Committee on Interstate and Foreign Commerce, Hearings on Natural Gas Amendments, 77th Cong., 1st sess. (July 1941), hereafter referred to as H.R. IFC 1941.

31. *Public Utilities Reports*, vol. 30, 247. Also see FPC dockets G-106 and G-119 in 2FPC29.

32. Also see "Federal Regulation of the Transportation and Sale of Natural Gas in Interstate Commerce," *AGA Proceedings* (1938), 126.

33. "Re: Tennessee Gas & Transmission Co," *Public Utilities Reports*.

34. Ibid., 137–39.

35. Ibid., 141.

36. Ibid.

37. Ibid., 145.

38. Tennessee Gas and Transmission Company, *Minutes*, October 15, 1941, Tenneco, Inc., History Collection.

39. Curtis B. Dall to John E. Buckingham, February 3, 1942, ibid.

40. H.R. IFC 1941, 1, 5.

41. Federal Power Commission, Staff Report on the Natural Gas Investigation, *Administration of the Certification Provisions of Section 7 of the Natural Gas Act* (Washington, DC: GPO, January 1947), 3.

42. H.R. IFC 1941, 5, 2–3.
43. Sanders, *Regulation of Natural Gas*, 52–53.
44. J. A. Krug to Curtis B. Dall, February 20, 1942, RG-138, box 147, folder 70-53, NARA.
45. Federal Power Commission, *Administration of the Certification Provisions of Section 7 of the Natural Gas Act*, 6.
46. Ibid., 9.
47. See FPC release, December 16, 1946, on unprecedented demand for natural gas, Texas Eastern Corp. History Collection.
48. As of April 24, 1942, the FPC and WPB agreed to coordinate their activities to meet war-related problems of the natural gas industry.
49. "The Ohio Fuel Gas Company and the Panhandle Eastern Pipeline Company," *Federal Power Commission Reports*, dockets G-408, G-410, vol. 4 (1942), 303.
50. Ibid., 305, 301–09.
51. "In the Matter of Tennessee Gas and Transmission Company," *Federal Power Commission Reports*, docket G-230, vol. 3 (July 5, 1943).
52. Harry L. Tower to Curtis B. Dall, December 8, 1942, Tenneco, Inc., History Collection.
53. "In the Matter of Tennessee Gas and Transmission Company," *Federal Power Commission Reports*, docket G-230, vol. 3 (July 5, 1943), 444.
54. Ibid., 445.
55. Ibid., 446.
56. Dall, *FDR*, 124–25.
57. Clyde Alexander, interview by Alan Dabney, July 3, 1962, Tenneco, Inc., History Collection.
58. Ibid. Also see "The History of the Tennessee Gas and Transmission Company," *Pipe Line News* (November 1944), 27–32.
59. E. H. Poe to Paul R. Taylor, September 4, 1943, Tenneco, Inc., History Collection.
60. Clyde Alexander, interview by Alan Dabney, July 3, 1962, ibid.
61. Joe Feagin, *Free Enterprise City* (New Brunswick, NJ: Rutgers University Press, 1988), 106–48.
62. "In the Matter of Tennessee Gas and Transmission Company," *Federal Power Commission Reports*, docket G-230, vol. 3 (September 24, 1943), 578, 579.
63. Tennessee Gas and Transmission Company, *Minutes*, September 20, 1943, Tenneco, Inc., History Collection.
64. The new board of directors included Clyde H. Alexander, Arthur D. Chilgren, Ray C. Fish, Charles F. Glore, Paul Kayser, Gardiner Symonds, and Richard Wagner.
65. "In the Matter of Tennessee Gas and Transmission Company," *Federal Power Commission Reports,* 574.
66. See index to minutes of the RFC, 1943, I-2, vol. 12, pt. I-B, 846, NARA.

67. Ray C. Fish, interview by Alan Dabney, April 30, 1963, Tenneco, Inc., History Collection. Also see Tennessee Gas and Transmission Company, *Prospectus*, December 11, 1945.
68. RFC Minutes, vol. 142, November 13, 1943, NARA.
69. Ibid., vol. 145, February 12, 1944, 302, NARA.
70. Richard Austin Smith, "Tennessee Gas Transmission vs. El Paso: They Play Rough in the Gas Business," *Fortune* (January 1965), 134.
71. Tennessee Gas and Transmission Company, *Prospectus*, December 11, 1945, Tenneco, Inc., History Collection.
72. Clyde Alexander, interview by Alan Dabney, July 3, 1962, ibid.
73. Frank H. Love, "Construction Features of the Tennessee Gas and Transmission Company Pipe Line," *Petroleum Engineer*, 16, no. 2 (November 1944), 121–44. Also see Tenneco, Incorporated, "Tenneco's First 35 Years," 1978.
74. Tennessee Gas and Transmission Company, *Annual Report* (1945), 2.
75. The $45,157,863 included a 102-½ charge for the loan including $57,863.01 in interest and a $1.1 million premium to the RFC.
76. Marshall McNeil, "Leesee of Big Inch Pipe Lines is War Baby of Curious History," *Washington Daily News* (December 9, 1946), 32.
77. Memorandum describing telephone call between Jesse H. Jones and Richard Wagner, RFC Minutes, vol. 155, December 21, 1944, NARA.
78. H. Malcolm Lovett, interview by Joseph A. Pratt and Christopher J. Castaneda, May 1, 1986, Oral History of the Houston Economy Collection, University of Houston.
79. "In the Matter of Tennessee Gas and Transmission Company and The Chicago Corporation," *Federal Power Commission Reports*, docket G-606, vol. 6 (May 28, 1947).
80. "In the Matter of Tennessee Gas and Transmission Company," *Federal Power Commission Reports*, docket G-621, vol. 4 (June 8, 1945), 294. Also see Defense Plant Corporation Minutes, roll 98, March 2, 1945, 114, NARA.
81. "In the Matter of Tennessee Gas and Transmission Company," *Federal Power Commission Reports*, docket G-621, vol. 4 (June 8, 1945), 295–96. Also see DPC Minutes, February 21, 1945, 960, and March 23, 1945, 1240, NARA.
82. "Tennessee Gas and Transmission Company," *FPCR*, 298–99.
83. The FPC granted a certificate for this pipeline on July 3, 1946, after Tennessee Gas resubmitted its application and showed that it could operate the extension economically. See "Tennessee Gas and Transmission Company," *FPCR*, docket G-660 (July 3, 1946).
84. Ibid., docket G-621, vol. 4 (June 8, 1945), 305–06.
85. Ibid., 302.
86. Ibid., 303, 304.

CHAPTER 4

1. See Arthur M. Johnson, *Petroleum Pipelines and Public Policy, 1906–1959* (Cambridge: Harvard University Press, 1967), and Christopher J. Cas-

taneda and Joseph A. Pratt, "The Limits of Strategy: A Strategic History of Texas Eastern Corporation" (unpublished manuscript).

2. Bill Murray, interview by Christopher J. Castaneda, December 12, 1990, Oral History of the Houston Economy, University of Houston.

3. U.S. Congress, House, Special Committee Investigating Petroleum Resources, Hearings on War Emergency Pipe-Line Systems and Other Petroleum Facilities, 79th Cong., 1st sess., November 15–17, 1945, hereafter referred to as H.R. PR 1945.

4. Johnson, *Petroleum Pipelines and Public Policy*, 326.

5. Ibid., 345.

6. Ibid., 343–45.

7. Sidney A. Swensrud, "A Study of the Possibility of Converting the Large Diameter War Emergency Pipe Lines to Natural Gas Service after the War," paper presented at the Petroleum Division of the American Institute of Mining & Metallurgical Engineers, New York City, February 24, 1944.

8. J. Howard Marshall, interview by Christopher J. Castaneda, December 13, 1988, Texas Eastern Corp. History Collection. Also see Otto Scott, *The Exception: The Story of Ashland Oil & Refining Company* (New York: McGraw Hill, 1968), 325.

9. Ralph K. Davies, deputy petroleum administrator, officially revoked petroleum directives 63 and 73 effective November 30, 1945. These directives provided for the wartime service of the pipelines. Also see Petroleum Administration for War release, November 14, 1945, Texas Eastern Corp. History Collection.

10. Reconstruction Finance Corporation release, September 14, 1945, Texas Eastern Corp. History Collection.

11. H.R. PR 1945, 3. See also Jesse J. Jones, *Fifty Billion Dollars: My Thirteen Years with the RFC* (New York: MacMillan, 1951), 402.

12. H.R. PR 1945, 24–25.

13. Ibid., 264–65.

14. Ibid., 346.

15. Minutes of the Petroleum Reserves Corporation, February 5, 1944, entry 222, 50, NARA.

16. H.R. PR 1945, 252–53.

17. Ibid., 261, 265.

18. Ibid., 385.

19. Ibid., 373.

20. Federal Power Commission, *Natural Gas Investigation—Nelson Lee Smith & Harrington Wimberly Report* (Washington, DC: GPO, 1948), hereafter referred to as Smith & Wimberly Report; Federal Power Commission, *Natural Gas Investigation—Leland Olds & Claude Draper Report* (Washington, DC: GPO, 1948), hereafter referred to as Olds & Draper Report.

21. Surplus Property Administration, *Government-Owned Pipelines*, Report of the Surplus Property Administration to Congress (Washington, DC: GPO, 1946), 2, 20.

22. U.S. Congress, House, *Second Interim Report of the Select Committee to Investigate Disposition of Surplus Property*, 79th Cong., 2nd sess., August 1946, 3–4.

23. *Statement by Committee on Post-War Disposal of Pipe Lines, Refineries and Tankers Subcommittee on Pipe Lines*, sent to Senator Joseph C. O'Mahoney, March 12, 1946, Texas Eastern Corp. History Collection.

24. *Oil & Gas Journal*, "WAA Proceeding Cautiously Disposing of Two Big Inch Pipelines" (June 8, 1946), 57.

25. Presidential Executive Order 9689, "Consolidation of Surplus Property Functions," February 1, 1946.

26. War Assets Administration, *Government Owned Pipelines: Report of the War Assets Administration to the Congress* (Washington, DC: GPO, December 18, 1946), 4, Texas Eastern Corp. History Collection.

27. AP wire service, July 30, 1946.

28. John P. Callahan, "Two of Sixteen Offers Propose Use of 'Big' and 'Little Inch' to Bring Supply from Southwest Planned," *New York Times*, August 4, 1946, sect. 3, 1–2.

29. War Assets Administration, *Transcript of Proceedings: Proposals to Buy or Lease the Big and Little Big Inch Pipe Lines* (Washington, DC: GPO, July 31, 1946).

30. War Assets Administration, press release, August 2, 1946, Texas Eastern Corp. History Collection.

31. "Former Big Names in Government are Involved in Bids for Pipelines," *Washington Daily News*, October 17, 1946.

32. James W. Hargrove, interview by Christopher J. Castaneda and Louis Marchiafava, March 31, 1988, Texas Eastern Corp. History Collection.

33. Dana Blankenhorn, "The Brown Brothers: From Mules to Millions as Houston's Contracting and Energy Giants," *Houston Business Journal*, March 19, 1979, sect. 2, 1–4. Also see Castaneda and Pratt, "Limits of Strategy."

34. Ronnie Dugger, *The Politician: The Life and Times of Lyndon Johnson* (New York: Norton, 1982), 282. Also see Robert Caro, *The Path to Power: The Years of Lyndon Johnson* (New York: Vintage, 1983), for background on the Brown brothers.

35. Dugger, *The Politician*, 282. Cook had worked at the SEC between 1937 and 1945 in a number of positions, including that of utilities analyst with the public utilities division, and he subsequently became executive assistant to United States attorney general Thomas Clark in 1945 and 1946.

36. Letter from Charles I. Francis to Judge J. A. Elkins, June 14, 1946, Texas Eastern Corp. History Collection. After Texas Eastern's successful bid, George Allen received substantial founders stock from Frank Andrews, an original stockholder. See Castaneda and Pratt, "Limits of Strategy."

37. Harold L. Ickes, " 'Uncle Jesse's' Bid for Oil Pipelines Seen Loaded Two Ways for Monopoly," *Evening Star* (October 2, 1946), A-11.

38. Harold I. Ickes to Charles I. Francis, October 18, 1946, Texas Eastern Corp. History Collection. Four years after this furor, George Butler joined the board of Texas Eastern.

39. War Assets Administration, press release, October 3, 1946, Texas Eastern Corp. History Collection.

40. E. Holley Poe to L. Gray Marshall, October 7, 1946, ibid.

41. Ibid.

42. See letter from F. J. Horne, Army-Navy Petroleum Board, to Robert Littlejohn, October 15, 1946, and J. S. Drug, secretary of the interior, to Robert Littlejohn, November 27, 1946, Texas Eastern Corp. History Collection.

43. See Elliot Taylor, "Thermally Thinking," *Gas* (November 1946), 21–22. Also see Marshall McNeil, "Former Big Names in Government Are Involved in Bids for Pipelines," *Washington Daily News* (October 17, 1946).

44. W. H. Leslie to George R. Brown, August 15, 1946, Texas Eastern Corp. History Collection.

45. "Big Inch Awards for Oil Seen Near; Poe Chides WAA," *Chicago Journal of Commerce* (October 23, 1946).

46. Harold L. Ickes, "WAA Administrator Urged to Sell 'Inch' Pipe Lines Now, and for Cash," *Evening Star* (October, 28, 1946); idem, "WAA Seen Yielding to Machinations of John L. Lewis on Natural Gas," ibid. (October 30, 1946); and see editorial, "The New Lewis Threat," *The News* (October 24, 1946).

47. "Lewis Home Uses Gas," *Washington Post* (December 6, 1946).

48. "Steelman Denies Ickes' Charge He Discussed Inch With Lewis," *Washington Post* (December 5, 1946).

49. J. Ross Gamble, "Notes Relative to Development on Big and Little Inch Pipe Lines," distributed to interested parties, November 8, 1946, Texas Eastern Corp. History Collection.

50. See War Assets Administration release, July 22, 1946, Texas Eastern Corp. Archives.

51. War Assets Administration, *Government-Owned Pipe Lines* (Washington, DC: GPO, December 18, 1946), 7.

52. U.S. Congress, House, *Report of the Select Committee to Investigate Disposition of Surplus Property*, House Resolution 385, 79th Cong., 2nd sess., 1947, 14.

53. Department of the Interior, press release, "Big Inch and Little Big Inch Conferences Held," November 30, 1946. Also see letters from Charles H. Smith, president Big Inch Oil, Inc., to General Robert M. Littlejohn, WAA, November 29, 30, 1946, Texas Eastern Corp. Archives.

54. U.S. Congress, House, Hearings, *Disposition of Surplus Property on House Resolution 385*, 79th Cong., 2nd sess., November 19–26, December 2–9, 1946, pt. 4, 2409.

55. G. H. McKay to WAA, October 23, 1946, Tenneco, Inc., History Collection.

56. R. G. Rice to the Department of the Interior, November 29, 30, 1946, ibid.

57. Letter (not dated) from War Assets Administration to Tennessee Gas in reference to PLE-1. (The company received the letter on December 3, 1946.) Tenneco, Inc., History Collection.

58. Federal Power Commission, *Statement on Natural Gas for the House Committee on Interstate and Foreign Commerce with Reference to H.J. Res. 2 on Disposition of the Big Inch and Little Big Inch Pipelines* (Washington, DC: GPO, January 23, 1947), 19.

59. War Assets Administration, *Government-Owned Pipe Lines*, 9.

60. August Belmont, interview by Christopher J. Castaneda and Joseph A. Pratt, June 14, 1988; James Ward Hargrove, interview by Christopher J. Castaneda and Louis Marchiafava, March 31, and April 13, 1988, Texas Eastern Corp. History Collection.

61. U.S. Congress, House, *Joint Resolution 2*, 80th Cong., 1st sess., January 3, 1947, Texas Eastern History Collection.

62. Statement of Congressman Francis E. Walter on Disposition of Big Inch Lines before House Committee on Interstate and Foreign Commerce, January 20, 1947, Texas Eastern History Collection.

63. R. G. Rice, interview by Alan Dabney, October 16, 1961, Tenneco, Inc., History Collection.

64. Ray Fish, interview by Alan Dabney, April 30, 1962, ibid.

65. Clyde Alexander, interview by Alan Dabney, July 3, 1962, ibid.

66. August Belmont, interview by Christopher J. Castaneda and Joseph A. Pratt, June 14, 1988, Texas Eastern Corp. History Collection.

67. Ibid.; August Belmont to Christopher Castaneda, February 6, 1989.

68. Winsor Watson, *History of the Texas Eastern Transmission Corporation* (no date). Also see interview of James Ward Hargrove by Christopher J. Castaneda, March 31, 1988, Texas Eastern History Collection.

69. August Belmont, interview by Christopher J. Castaneda and Joseph A. Pratt, June 14, 1988, Texas Eastern Corp. History Collection.

70. War Assets Administration, *Transcript of Proceedings: Proposals for Purchase of War Emergency Pipe Lines Commonly Known as Big and Little Big Inch Pipe Lines* (Washington, DC: GPO, February 10, 1947), ibid.

CHAPTER 5

1. Clark, *Energy and the Federal Government*, xxi, 1–2.

2. H.R. 4816 was introduced by Mr. Cole of Maryland.

3. *Congressional Record*, June 5, 1941, 4783.

4. John Loomis to James T. Daly, May 20, 1946, Texas Eastern Corp. History Collection.

5. "Pennsylvania Holds Key to Use of Big and Little Inch for Gas," *Philadelphia Bulletin* (November 27, 1946).

6. "Bidder Sees No Bar to Gas Lines," *Evening Bulletin* (February 11, 1947).

7. Castaneda and Pratt, "Limits of Strategy," 110.

8. See "Memo on questions propounded to Mr. J. Ross Gamble regarding the proposal of Texas Eastern Transmission Corporation for the purchase of the Big Inch and Little Big Inch Pipe Lines" and Gamble's response of February 20, 1947, Texas Eastern Corp. Archives.

9. Robert H. Poe, interview by Christopher J. Castaneda and John O. King, June 20, 1988, ibid.

10. J. Howard Marshall, interview by Christopher J. Castaneda, December 13, 1988, ibid.

11. James W. Hargrove, interview by Christopher J. Castaneda and Louis Marchiafava, March 31, 1988, Texas Eastern Corp. History Collection.

12. John K. Weiss, "How Bankers Make Millions," *PM Daily*, 3, no. 126 (November 11, 1947).

13. "Bidder Sees No Bar to Gas Line," *Evening Bulletin* (February 11, 1947).

14. Federal Power Commission, "Daily News Digest," 21, no. 35 (February 19, 1947), 2.

15. Charles I. Francis to Judge J. A. Elkins, June 14, 1946, Texas Eastern Corp. History Collection.

16. Letter from August Belmont to Christopher J. Castaneda, March 21, 1990. According to company stock records, George Allen later received a sizable stock transfer from Frank Andrews.

17. "Duff Doubts Pipeline Curb," *Philadelphia Bulletin* (February 15, 1947).

18. August Belmont, interview by Christopher J. Castaneda and Joseph A. Pratt, June 14, 1988, Texas Eastern Corp. Archives.

19. General Robert Littlejohn to Tennessee Gas Transmission Company, March 8, 1947, Texas Eastern Corp. History Collection.

20. Ibid.

21. Minutes of the meeting, March 3, 1947, Tenneco, Inc., History Collection.

22. "A Need for Speed," *GAS* (March 1947).

23. Ibid.

24. J. A. Walter to Leon Fuquay, May 6, 1947, Texas Eastern Corp. History Collection.

25. "Texas Eastern Chairman Criticizes Gas Barrier," *Oil and Gas Journal* (May 24, 1947), 126.

26. Charles I. Francis to E. R. Cunningham, May 24, 1957, Texas Eastern Corp. History Collection.

27. John A. Danaher to Charles I. Francis, February 12, 1947, includes marginalia from Charles I. Francis, ibid.

28. U.S. Congress, House, Committee on Interstate and Foreign Commerce, Hearings on Amendments to the Natural Gas Act, 80th Cong., 1st sess., April–May 1947, hereafter referred to as H.R. IFC 1947.

29. Vietor, *Energy Policy in America since 1945*, 76–8.

30. H.R. IFC 1947, 380, 423.

31. "Eminent Domain for Gas Pipe Lines Provided in Senate Bill," *Oil and Gas Journal* (April 12, 1947), 55.
32. Charles Francis to August Belmont, May 1, 1947, Texas Eastern Corp. History Collection.
33. Petition, Natural Gas Requirements of the Philadelphia Gas Works Company, "In the Matter of Texas Eastern Transmission Corporation," docket G-880 (June 1947), 1, ibid.
34. Nicholas B. Wainwright, *History of the Philadelphia Electric Company: 1881–1961*. (Philadelphia: Philadelphia Electric Company, 1961), 278, 280.
35. Ibid., 279.
36. Ibid., 299.
37. Petition of the City of Philadelphia, a Municipal Corporation of the First Class of the Commonwealth of Pennsylvania, for Leave to Intervene as a Party, "In the Matter of Texas Eastern Transmission Corporation," FPC docket G-880, signed by Bernard Samuel, mayor of Philadelphia, June 27, 1947, Texas Eastern Corp. History Collection.
38. "In the Matter of Texas Eastern Transmission Corporation," *Federal Power Commission Reports*, docket G-880, vol. 6 (October 10, 1947), 148–75.
39. "State Now Backs Importation of Natural Gas," *Philadelphia Bulletin* (July 11, 1947).
40. Jack Head, also a member of Vinson, Elkins, Weems & Francis, would later succeed Francis as Texas Eastern's general counsel.
41. Federal Power Commission, Hearings, "In the Matter of Texas Eastern Transmission Corporation," docket G-880 (August 25, 1947), 4609–10, Texas Eastern Corp. History Collection.
42. Ibid., 4611.
43. "In the Matter of Texas Eastern Transmission Corporation," *Federal Power Commission Reports*, docket G-880, vol. 6 (October 10, 1947), 109.
44. "Duff Backs Philadelphia Plea for Natural Gas Supply to Avert Rate Increase," *Philadelphia Inquirer* (Sept. 25, 1947), 1.
45. Federal Power Commission, Hearings, "In the Matter of Texas Eastern Transmission Corporation," docket G-880, 2299, Texas Eastern Corp. History Collection.
46. "Many Join City's Drive to Get Natural Gas," *Philadelphia Inquirer* (September 25, 1947).
47. "New Battle Opens on Big, Little Inch," *New York Times* (September 26, 1947).
48. "In the Matter of Texas Eastern Transmission Corporation," *Federal Power Commission Reports*, vol. 6 (Washington, DC: GPO, 1949), 167.
49. "Flow of Natural Gas Expected to Start Early Next Year," *Philadelphia Inquirer* (October 13, 1947).
50. See John W. Welker, "Fair Profit?" *Harvard Business Review* 26 (March 1948), 207–15, and Joseph Stagg Lawrence, "Profits and Progress," ibid. (July 1948), 480–91.

51. August Belmont, interview by Christopher J. Castaneda and Joseph A. Pratt, June 14, 1988, Texas Eastern Corp. Archives. Also see Texas Eastern Minutes, 1947.
52. J. Ross Gamble to Charles I. Francis, January 14, 1948, Texas Eastern Corp. History Collection.
53. R. H. Hargrove to J. Ross Gamble, February 2, 1958, ibid.
54. Leon M. Fuquay to J. Ross Gamble, March 10, 1948, ibid.
55. Petition of the Philadelphia Gas Works Company for a Declaratory Order to Remove Uncertainty, June 1, 1948, ibid.
56. "In the Matter of Texas Eastern Transmission Corporation," docket G-880, answer of Texas Eastern Transmission Corporation to the Petition of the Philadelphia Gas Works Company for a Declaratory Order to Remove Uncertainty, Before the FPC of the U.S.A., dated June 10, 1989, ibid.
57. Wainwright, *History of the Philadelphia Electric Company*, 278.
58. Ibid., 319.

CHAPTER 6

1. Ralph E. Davis, *Stories of Natural Gas* (Ralph E. Davis, 1964), 44, 45.
2. Ray C. Fish, interview by Alan Dabney, April 30, 1962, Tenneco, Inc., History Collection.
3. Claude A. Williams, "The Story of Transcontinental," *Oil and Gas Journal* (May 4, 1950), 81–83.
4. "Government Utility Happenings," *Public Utilities Fortnightly*, 37 (April 25, 1946), 556–57. Also see Williams, "Story of Transcontinental," and Ray C. Fish, interview by Alan L. Dabney, April 30, 1962. Note that the company subsequently changed its name to Transcontinental Gas Pipe Line Company.
5. Notice of Amended Application for a Certificate of Public Convenience and Necessity, "In the Matter of Trans-Continental Gas Pipe Line Company, Inc.," docket G-704 (January 10, 1947), Texas Eastern Corp. History Collection.
6. Ibid.
7. Alfred C. Glassell, Jr., interview by Christopher J. Castaneda, April 26, 1990, Oral History of the Houston Economy, University of Houston.
8. Note that this amount was $1 million less than Transcontinental's second bid for the Inch Lines.
9. From Notice of Amended Application, statement issued by Leon Fuquay, secretary of the FPC, on January 10, 1947, docket G-704. The utilities were Consolidated Edison Co., Brooklyn Union Gas Co., Public Service Corp., Philadelphia Gas Works Co., Philadelphia Electric Co., Consolidated Gas Electric & Power Co., Delaware Power & Light Co., and Elizabeth Consolidated Gas Co., Texas Eastern Corp. History Collection.
10. Ibid.

11. Thomas O. Waage, "New York Area Held Sure to Get Natural Gas by Pipelines Soon," *New York Journal of Commerce* (December 23, 1946).
12. Petition by Eastern States Retail Solid Fuel Conference for Leave to Intervene, docket G-704 (January 30, 1947), Texas Eastern Corp. History Collection.
13. Petition by Anthracite Institute for Leave to Intervene, "In the Matter of Application by Trans-Continental Gas Pipe Line Company, Inc. for Certificate of Public Convenience and Necessity," January 15, 1947, docket G-704, ibid.
14. Joint Petition for Leave to Intervene, "In the Matter of Trans-Continental Gas Pipe Line Company, Inc.," docket G-704 (January 28, 1947), ibid.
15. John Bauer to Honorable Nelson Lee Smith, December 30, 1946. Includes accompanying correspondence from John Bauer to the FPC, December 30, 1946, ibid.; Protest and Petition to Intervene of Big Inch Natural Gas Transmission Company, docket G-704, ibid.; Petition to Intervene of Southern Natural Gas Company, "In the Matter of the Application of Trans-Continental Gas Pipe Line Company, Inc.," docket G-704 (January 30, 1947), ibid.
16. Claude Williams to the FPC, February 17, 1947, ibid.
17. Leon Fuquay to Claude A. Williams, February 24, 1947, ibid.
18. "Trans-Continental Gas Pipe Line Co. Urges Texas-N.Y. Gas Line Hearing by F.P.C.," *FPC Daily News Digest*, 21, no. 36 (February 20, 1947), 2.
19. "N.Y. State Natural Gas Corp. Plan for Exports Opposed," *Wall Street Journal* (February 26, 1947).
20. Richard W. Hooley, *Financing the Natural Gas Industry* (New York: Columbia University Press, 1961), 126.
21. Paul L. Howell and Ira Royal Hart, "The Promoting and Financing of Transcontinental Gas Pipe Line Corporation," *Journal of Finance*, 61, no. 3 (September 1951), 314.
22. Petition to Intervene of Texas Eastern Transmission Corporation, "In the Matter of Trans-Continental Gas Pipe Line Company, Inc.," docket G-704 (May 29, 1947), Texas Eastern Corp. History Collection.
23. H. Malcolm Lovett, interview by Joseph A. Pratt, May 20, 1985, Oral History of the Houston Economy, University of Houston.
24. Reginald H. Hargrove to W. E. Bolte, June 11, 1947, Texas Eastern Corp. History Collection.
25. Federal Power Commission, Hearings, "In the Matter of Texas Eastern Transmission Corporation," docket G-880, 4958–66, ibid.
26. "In the Matter of Trans-Continental Gas Pipe Line Company, Inc.," *Federal Power Commission Reports*, docket G-704, vol. 7 (May 29, 1948), 28–30.
27. Ibid., 30.
28. Ibid., 44.
29. Ibid., 45.
30. Ibid., 32.

31. See State of New York Public Service Commission, *Utility Regulatory Bodies in New York State: 1855–1953* (Albany, 1953).
32. New York Public Service Commission, *Annual Report* (1948), 95.
33. "In the Matter of Trans-Continental Gas Pipe Line Company, Inc.," *Federal Power Commission Reports*, docket G-704, vol. 7 (May 29, 1948), 32–34.
34. Ibid., 36, 43.
35. Ibid., 43.
36. Ibid.
37. Howell and Hart, "Promoting and Financing of Transcontinental Gas Pipe Line Corporation," 315.
38. Petition to Amend Prior Order and Application for Certificate of Public Convenience and Necessity, "In the Matters of Trans-Continental Gas Pipe Line Co., Inc. and Trans-Continental Gas Pipe Line Corporation: G-704, G-1143," dockets G-704, G-1143 (November 17, 1948), Texas Eastern Corp. History Collection.
39. Transcontinental Gas Pipe Line Corporation, *Annual Report, 1949.*
40. "Natural Gas — Woosh," *Fortune*, 40, no. 114 (December 1949), 199. Also see Texas Eastern Transmission Corporation, *Annual Report* (1949), 6–7.
41. Wainwright, *History of the Philadelphia Electric Company*, 320–21.
42. Joseph A. Pratt, *A Managerial History of Consolidated Edison, 1936–1981* (Consolidated Edison Company of New York, 1988), 166.
43. Martin Toscan Bennet, "The 'Safest Inch,'" *Engineering News-Record* (September 14, 1950).
44. "Texas Gas Comes Through," *New York Times* (January 22, 1951).
45. Pratt, *Managerial History of Consolidated Edison*, 163.
46. Ibid., 169.
47. Ibid., 171.
48. David F. Grozier, "The Brooklyn Union Natural Gas Conversion: Biggest Changeover in the World," *Gas Age* (January 1, 1953), 31.
49. Ibid., 33.
50. Ibid., 32.
51. Ibid., 33–34.
52. Ibid., 30.

CHAPTER 7

1. The New England states are Connecticut, Maine, Massachusetts, New Hampshire, Rhode Island, and Vermont.
2. Terry, "Natural Gas for the Northeastern Seaboard."
3. Alfred C. Glassell, Jr., interview by Christopher J. Castaneda, April 26, 1990, Oral History of the Houston Economy, University of Houston.
4. "Natural Gas — Woosh," 204.
5. "Businessmen in the News: Principals in Battle for New England's Gas Market," *Fortune* (January 1953), 37.

6. "Texas in New England," *Forbes*, 70, no. 11 (December 15, 1942), 14.

7. John Osborne, "A Brawling, Bawling Industry," *Life* (March 10, 1952), 101.

8. "Texas in New England," 15.

9. H. Malcolm Lovett, interview by Joseph A. Pratt, May 20, 1985, Oral History of the Houston Economy, University of Houston.

10. Tennessee Gas Transmission Company, *Annual Report* (1950), 11.

11. "Natural Gas for New England," *A Debate Held Under the Auspices of the New England Council at Providence, Rhode Island, March 19, 1948* (reprinted by Fuels Research Council, Inc., Washington, DC), 2, Texas Eastern Corp. History collection.

12. Ibid.

13. Ibid., 5.

14. Ibid., 11.

15. R. H. Hargrove to the Honorable Philip J. Philbin, February 6, 1948, Texas Eastern Corp. History Collection.

16. Texas Eastern Transmission Corporation, *Annual Report* (1948), 22.

17. Robert M. Poe, interview by Christopher J. Castaneda and John O. King, June 20, 1988. Also see "Natural Gas — Woosh," 105.

18. "Natural Gas — Woosh," 105.

19. Texas Eastern Transmission Corporation, *Annual Report* (1949), 16.

20. The FPC approved the Kosciusko line application on February 26, 1951.

21. Texas Eastern Transmission Corporation, *Annual Report* (1949), 16. Also see idem, *Annual Report* (1951), 13.

22. Tennessee Gas Transmission Company, *Annual Report* (1949), 23.

23. Algonquin Calendar, 6, Texas Eastern Corp. History Collection.

24. "In the Matter of Northeastern Gas Transmission Company and Algonquin Gas Transmission Co.," *Federal Power Commission Reports*, dockets G-1568, G-1319, vol. 10 (March 1, 1951), 87.

25. Speech by Gardiner Symonds, "Natural Gas for New England," before the 96th Quarterly Meeting of the New England Council, Harbor, Maine, September 16, 1949, Texas Eastern Corp. History Collection.

26. Ibid.

27. "Natural Gas — Whoosh," 204. Also see "Battle for New England," *Time*, 60, no. 23 (December 8, 1952), 53.

28. Final Ownership: EG&FA: 36.8%; NEGEA: 34.5%; Providence Gas Company: 0.7%; Texas Eastern: 28%.

29. R. H. Hargrove, "Natural Gas and New England," address to New England Council, Polar Springs, Maine, September 16, 1950, 14, Texas Eastern Corp. History Collection.

30. F. D. Campbell to George Naff, February 13, 1953, Texas Eastern Corp. History Collection.

31. Report, "Oil Opposition to Natural Gas," by Ross McKee for Winsor H. Watson, February 1950, ibid.

32. R. H. Hargrove, "Natural Gas and New England," address to the New England Council, Poland Springs, Maine, September 16, 1950, ibid.

33. "In the Matter of Northeastern Gas Transmission Co. and Algonquin Gas Transmission Co.," *Federal Power Commission Reports*, Dockets G-1568, G-1319, vols. 10, 12 (March 1, 1951), 92.

34. John F. Rich to Gardiner Symonds, October 9, 1950, Texas Eastern Corp. History Collection.

35. Telegram from Gardiner Symonds to J. F. Rich, October 9, 1950, ibid.

36. Osborne, "Brawling, Bawling Industry," 104, 107.

37. "In the Matter of Northeastern Gas Transmission Co. and Algonquin Gas Transmission Co.," *Federal Power Commission Reports*, 94–96.

38. Tennessee Gas Transmission Company, press release, January 31, 1951, Tenneco, Inc., History Collection.

39. "Opinion 259," *Federal Power Commission Reports* (August 1953), 2–4. This opinion concluding the litigation in the Algonguin-Northeastern matter reviews the extensive legal maneuvering from 1950 through 1953.

40. "In the Matter of Northeastern Gas Transmission Co. and Algonquin Gas Transmission Co.," *Federal Power Commission Reports*, 97.

41. Ibid., 99, 100.

42. "In the Matter of Northeastern Gas Transmission Co.," *Federal Power Commission Reports*, Docket G-1267, vol. 10 (March 21, 1951), 137–40.

43. Tennessee Gas Transmission Company, press release, November 24, 1951, Tenneco, Inc., History Collection.

44. Algonquin Calendar, 4, Texas Eastern Corp. History Collection. Also see "Battle for New England," 52.

45. Northeastern Gas Transmission Company, *Minute Book*, November 12, 1951, Tenneco, Inc., History Collection.

46. Osborne, "Brawling, Bawling Industry," 108.

47. Algonquin Calendar, 29, Texas Eastern Corp. History Collection.

48. John Mackin, interview by Joseph A. Pratt, Kenneth Lipartito, and Christopher Castaneda, November 19, 1986, Oral History of the Houston Economy, University of Houston. Also see "Foes of Pipelines Seek Court Help," *New York Times* (October 23, 1952).

49. "Battle for New England," 53.

50. Texas Eastern Transmission Corporation, *Annual Report* (1953), 3.

51. Osborne, "Brawling, Bawling Industry," 107–08.

52. Ibid., 108.

53. Ibid.

54. Algonquin Calendar, 29, Texas Eastern Corp. History Collection.

55. *Boston Herald* (November 7, 1952).

56. Algonquin Calendar, 30, 32, Texas Eastern Corp. History Collection.

57. Advertisements in the *Boston Herald* in 1953 (January 12): "A Plain Statement about Natural Gas in New England: The Fact is Northeastern Can Bring It to You Cheaper"; (January 14): "Who Built the Natural Gas

Roadblock? Why Was It Built? Will It Be Removed?"; (January 16): "From Every Angle Northeastern Can Provide All New England with the Best Natural Gas Service."

58. Editorial, "Rebuttal on Gas," *Boston Herald* (January 18, 1953).

59. Algonquin Calendar, 42, 45, Texas Eastern Corp. History Collection.

60. Tennessee Gas and other interested parties to the FPC, July 1, 1953, Tenneco, Inc., History Collection.

61. Federal Power Commission, "Opinion and Order Issuing Certificate of Public Convenience and Necessity," no. 259 (July 1953), 5.

62. "Comments of Intervenor Trans-Canada Pipe Line Ltd. in opp. to approval of the Commission of the Terms of Settlement Proposed," July 3, 1953, 6, Texas Eastern Corp. History Collection.

63. "Time Clock," *Time*, 62, no. 79 (August 17, 1953).

64. "FPC Lets Natural Gas Company Ship to Canada," *Business Week*, 1253, no. 30 (September 5, 1953).

65. Algonquin Gas Transmission Company, *Minutes*, January 15, 1954, Texas Eastern Corp. History Collection.

CHAPTER 8

1. Claude A. Williams, "Natural Gas Transmission and Its Future," address to the National Federation of Financial Analysts Societies Convention, March 7, 1951, Texas Eastern Corp. History Collection.

2. Notes from December 16, 1953, conference, Tenneco, Inc., History Collection.

3. R. H. Hargrove, "Where Do We Go From Here?" address before the Eastern Group of the Investment Bankers of America, Philadelphia, April 13, 1951, 6, Texas Eastern Corp. History Collection.

4. August Belmont to Christopher Castaneda, March 21, 1990.

Index